The Practical Imperialist

Sources For African History

VOLUME 4

The Practical Imperialist

Letters from a Danish Planter in German East Africa 1888-1906

Edited by

Jane L. Parpart
and
Marianne Rostgaard

BRILL
LEIDEN · BOSTON
2006

This book is printed on acid-free paper.

Library of Congress Cataloging-in-Publication Data

The practical imperialist : letters from a Danish planter in German East Africa 1888-1906 / edited by Jane L. Parpart and Marianne Rostgaard.
p. cm. — (Sources for African history, 1570-8721 ; v. 4)
Includes bibliographical references and index.
ISBN 900414742X (alk. paper)
1. Lautherborn, Christian—Correspondence. 2. Farms—Africa. 3. Germany—Colonies—History. 4. Tanzania—History.

JV2011+ 2005

2005058140

ISSN 1570–8721
ISBN 90 04 14742 X

PRINTED IN THE NETHERLANDS

CONTENTS

LIST OF ILLUSTRATIONS

2. No text and no date. Maybe the long house covered with palm leaves, used for workers quarters, mentioned in this letter 13th November 1891.
3. No text and no date. On the shore coral stone. See description of mortar making in chapter 6.
4. No text and no date. The workers are fabricating bricks.
5. No text and no date. In the foreground 'makutis'.
6. 'At Kikogwe. Herr Nietzch, Frau Hollmann and I'.

7a and 7. No text. Maybe the European colony in the Pangani area on a visit to Kikogwe, mentioned in letter 8th October 1891.

7. CL stands in the front row, number three from the right with what looks like a leash for the dog in his hand.
8. No text, but there were no railroads to Pangani or Bagamoyo in the 1890's so it must be Tanga.
9. No text and no date.

Chapter 8

1. 'My coffee-plantation'.

2 and 2b. No text and no date—The Pangani river close to Kikogwe.

3. No text and no date, but the photo must be from one of CL's plantations, most probably Kikogwe.
4. 'Lay down of the foundation stone for the sugar-factory 27th January 1899'.
5. No text and no date, but it may be the Christmas party mentioned in the letter 4th April, 1898.
6. No text and no date, but obviously an outing.

Chapter 9

1. No text and no date. CL with two of his assistants and a locomobile.
2. 'My donkey-cart'.
3. 'My house in Mwera seen from the side'.
4. Touristing in Egypt. CL is riding the donkey.
5. A picture of the Carlsen-family taken in Lønstrup when CL was on 'home-leave', summer 1900. Vilhelm Carlsen stands in the back row (without hat). Caroline is sitting in the second row, number two from the right (with a small black hat). CL sits in the second row on the left side. The rest of the flock is the Carlsen children—the oldest of them together with their spouses.

FOREWORD

Mogens Thøgersen, Director, Vendsyssel Historical Museum
and Archives

Vendsyssel Historical Museum in Hjørring is a local museum in Northern Jutland, Denmark, founded in 1889. As other local museums the main task is to preserve the local or regional heritage. This more or less begs the question why a collection of weapons and other artefacts from German East Africa, today's Tanzania, ended up in our museum and why it was accepted by the curators of the museum?

A very pragmatic reason was that the museum in the first years was run by a group of interested citizens, not professional museum-people of today that may have asked questions like whether it was our job to preserve these artefacts or they should go into another collection. The leading figure and first foreman of the group of enthusiasts who founded the museum and constituted its first board was the local dentist Lønborg Friis. Christian Lautherborn's brother in law, Vilhelm Carlsen, who played an essential role in social and cultural life of the rather small provincial town was also among the backers and of course new Lønborg Friis, and Christian Lautherborn knew him too. It was therefore a rather natural thought for this group of people that the artefacts Christian Lautherborn had sent from East Africa should go into the museum.

The board, however, also answered the question why in an announcement of the acceptance of the first part of the artefacts collected by Christian Lautherborn in Vendsyssel Tidende in 1892: 'These artefacts do not directly tell us about Danish history. On the other hand it is important for the museum to collect such items because they can illuminate the life of our forefathers who once used similar weapons.'

The acceptance of the collection was in accord with the once very popular stage-thinking, where each society was supposed to go through a number of stages and thus move from a less to a more civilised stage. The artefacts from East Africa were thus supposed to supplement the collection of items from pre-historic Denmark. The arte-facts from East Africa, however, were never displayed according to the original intent. When displayed for the first time in the 1960's it was to add an exotic touch to an exhibition of aid to the third world.

The artefacts were taken out of their hiding for the second time in 2001 in order to make an exhibit which focused on German East Africa in Christian Lautherborn's time, seen through his letters and photographs. The exhibition was based on the work done by Anna Marie Knudsen, first as a trainee later as a temporary employee at the museum.

The exhibition was shown during the summer of 2001, where Dr. Eginald Mihanjo, Professor , Department of History, University of Dar es Salaam, and secretary general of the Historical Association of Tanzania, happened to come by Aalborg University, serving as a member of a Ph.D. evaluation committee at the university in Aalborg. It was of course an obvious idea for Marianne Rostgaard and Anna Marie Knudsen to inform him about the exhibition and escort him to Hjørring.

Very little material from the early part of the German colonial period can today be found in Tanzania and thus be used to study the social and cultural history of early colonialism in Tanzania, and what has survived has normally survived in archives in Germany—both linguistic and economic difficulties will make it hard for a student in Tanzania today to get a chance to study the original sources. Dr. Eginald Mihanjo therefore showed a keen interest in having the letters translated and the exhibition transferred to Dar es Salaam. As the first part of this scheme was easier to realise than the last it was decided to start with the translation of the letters into English. Dr. Mihanjo would thereafter try to get the letters translated into Kiswahili as well.

Our museum was happy to receive a small but nevertheless very welcome donation from Danida which helped us to finance translation of the letters, copying of photos etc. We are also happy to see that the translated letters are now going into print. In this way the little bits and pieces of a part of the history of Denmark, Germany and Tanzania now has the chance of being weaved together to what is in fact part a common history. Christian Lautherborn was a Dane and it significantly impacted his thinking and doings, but the history, or parts of history told through his letters are also German and Tanzanian. The real story can therefore only be told as a transnational history of which we are proud to have contributed one piece in a vast jig-saw puzzle.

Hjørring 10th April 2005
Mogens Thøgersen

FOREWORD BY THE EDITORS

Christian Lautherborn is not a well-known historical person, but one of the many rather anonymous agents of imperial colonisation of Africa in the late 1800's and early 1900's. Christian Lautherborn went to East Africa as a plantation manager hired by the German East African Company (DOAG) in 1888, at twenty-eight years old. He wrote letters home to his family until his death in 1906, first and foremost to his sister Caroline Carlsen. Thanks to Caroline Carlsen the letters were preserved and ended up in The Royal Library in Copenhagen. The letters, translated into English, appear here for the first time in a printed edition.

The letters are fuller and longer during the first years of Lautherborn's stay in Tanzania; the many new impressions and experiences had to be related to somebody. Moreover, Caroline's husband, the editor of the *Vendsyssel Tidende*, the Hjørring newspaper of the period, urged Lautherborn to send letters about his adventures that would interest the North Jutland reading public. Hence, many of especially the earlier letters were published in that newspaper. As the years went on and his life in East Africa became more of an everyday experience, the letters were fewer and shorter, but they are, in general, very instructive and substantial. Christian Lautherborn is a good narrator and a perceptive observer. This alone might have justified an edition of Chr. Lautherborn's letters from Africa. But besides, we do not have many eye witness accounts related, not from the point of view of the missionary, adventurer or colonial officer, but from the everyday viewpoint of a manager of colonialism. Christian Lautherborn writes from the point of view of a practical imperialist, bent on introducing modern agricultural methods and producing goods for a profit. Indeed, he was the first to prove that sisal could be grown commercially in German East Africa (GEA)—a crop that was to become the region's leading export. Yet, Lautherborn also provides an interesting double vision—as a Dane working for a German institution, his letters reveal some of the tensions around class, national identity and culture that permeated the colonial enterprise. Much has been written about the colonial-indigenous encounter; these letters remind us that relations within the colonial community were often complicated and power-laden as well. Finally Lautherborn writes in a period that is poorly documented, particularly in English, as many original sources were destroyed during World War I. All of these reasons, along with his careful observations of daily life and his interest in African attitudes and perceptions (enhanced by his early fluency in Swahili) provide ample justification for making these letters from Africa known and accessible to a wider audience.

We have had the privilege to edit the letters, but numerous people have helped us along the way. One person, however, deserves thanks more than all others—a graduate student, Anna Marie Knudsen, Aalborg University, who discovered the letters during a trainee-ship at Vendsyssel Historiske Museum in Hjørring. Before Anna Marie enrolled as a history major, she and her husband had worked more than ten years in Tanzania. This was the reason why the director at the museum in Hjørring asked Anna Marie if she would be interested in taking a look at their East African collection, stored away in a corner of the museum's magazines. Anna Marie started to dig into the history of why artefacts from German East Africa had ended up in a provincial museum in Northern Jutland, Denmark. It is thanks to her dedication and ability as a detective that the letters and photos were retrieved and, almost like a film being developed into negatives in a dark room, a fascinating story began to surface.

Anna Marie moved to Greenland to take up a teaching job and left the letters and substantial additional information about Christian Lautherborn for us to continue work on, to bring, at least when we speak of Denmark, an almost forgotten part of history back to memory. In addition we would like to thank the staff at the museum in Hjørring for their willingness to provide information and help us along the way. The photos in the book have been reproduced by courtesy of Niels Carlsen and Vendsyssel Historiske Museum.

Special thanks also goes to Steven Fabian, a PhD student from Dalhousie University in Halifax, Canada, where Jane Parpart is a professor. Steven is writing a thesis on the history of Bagamoyo, and has provided important data and insights from the field. Thanks to the wonders of email, we have been able to send out questions to Steven that could then be checked in the Tanzanian National Archives—a rare treat and an important contribution to our understanding of the letters and the accuracy of the footnotes. We also are grateful to helpful comments from and the scholarship of: Jan-Georg Deutsch (Oxford University), Stephen Rockel (University of Toronto), Jim Brennan (SOAS, University of London), Thaddeus Sunseri (Colorado State University and Juhani Koponen (University of Helsinki). As non-specialists on Tanzania, we have been overwhelmed by the support and interest of Tanzanian specialists in this project and hope that the letters and background chapters reflect their generosity and assistance. Chapter two has also been presented at a number of seminars, where insightful and challenging questions have helped to hone our thinking. Any shortcomings in the editorial notes of the letters and the background chapters reflect our own limitations and not those of our many helpful advisors.

Of course, without libraries and archives, a work such as this would not be possible. Bits and pieces of information have been gathered from The Royal Library in Copenhagen, Stadsarkivet (The City Archives) in Aalborg, Vendsyssel Historiske Museum og Arkiv in Hjørring and the library at Aalborg University. Thanks also goes to the staff and the director of the Institute of Commonwealth Studies, School of Advanced Studies, London

University for generously hosting us the three times we met for a couple of days to coordinate the editing and discuss the introductory chapters. For a professor from Aalborg University, Denmark and from Dalhousie University, Nova Scotia, Canada, London was the ideal place to meet. Moreover, the libraries at the School of Oriental and African Studies (University of London) and Institute of Commonwealth Studies were well suited to our search for background information, and have been extensively used. We are grateful to all of these institutions for their assistance.

Aalborg and Dalhousie Universities, April 2005
Jane L. Parpart (Dalhousie) and Marianne Rostgaard (Aalborg)

ZIGUA
TANGA
Tanga
Mombo
Soni
Gombelo
Amani
Korogwe
Muheza
Pongwe
Handeni
Mgera
Kwinji
Pangani
Falls
Mwera
Msangasi
Mkwaja
I A
Njoro
NGURU
Kibati
Mziha
Mligasi
Sadani
Kanga
Matale
Mhonda
Mbwewe
Turiani
Wami
Mvomero
Mandera
Kidete
Rusako
Dakawa
Lugoba
Bagamoyo
Chalinze
Msowero
Mkata
Rudewa
Mbwawa
Kerege
Kimamba
Ngerengere
Kidugallo
Ruvu
Soga
Kilosa
Morogoro
Kisarawe
Pugu
DAR ES SALAAM
Uluguru Mts
Mkuyuni
Mzenga
Kola
Kimbiji
Mkuranga
Mpakani
Matombo
Mvuha
Maneromango
Bigwa
Kuruti
MIKUMI
Dakawa
Dutumi
Ruvu
Kisaki
Kisangire
Kisiju
NAT. PARK
COAST
Kilengwe
Kwa Mhinda
Kibiti
Msindaji
Kidodi
Mtegere
Ngulakula
Kikale
Kidatu
Ruaha
Rufiji
Zombe
Mpanganya
Kilindoni
Kiberege
Rapids
Rusende
L. Utenge
Utete
Pemba Channel
Ras Kiuyu
Wete
PEMBA I.
Chake Chake
Mkoani
Kengeja
Ras Upembe
Panza I.
Ras Nungwe
Mkokotoni
ZANZIBAR
Chwaka
ZANZIBAR I.
Zanzibar Channel
Fumba
Makunduchi
Kizimkazi
Ras Mkumbi
Bweni
Kirongwe
Mafia Channel
Mafia I.
Chole
I N
O

NOTE ON EDITING

The original letters in Danish have been translated into English. The only Danish words that have been kept are names of places and people in Denmark and Danish measures. CL's spelling has been preserved unless it was obviously a spelling error.

Regarding names of people and places in German or Swahili, either in Germany or in German East Africa we have consulted a number of reference works. Names of places, etc. are rendered in the conventional or most commonly used form. German or Swahili words are italicised in the text. Words are only underlined if they were underlined too in the original letters in Danish. The same goes for parentheses, whereas brackets have been inserted by the editors for short explanations, etc.

A few letters have been omitted; we have noted when and shortly retold the contents of these letters which mostly deal with family business home in Denmark. Most of the original, hand-written letters have survived and are now in the Royal Danish Library in Copenhagen. Some letters, however, only exist today in the version in which they were printed in *Vendsyssel Tidende*. We have a few letters in both a newspaper and the original hand-written version and have thus been able to compare. The newspaper editor (CL's brother-in-law Vilhelm Carlsen) has made minor corrections so that the text in the newspaper version is sometimes more clear than in the original letter, but the changes are always insubstantial. Basically, the letters were printed in the newspaper as they were apart from greetings to the family, etc.

We have indicated whether the letter is the original or the newspaper version. Where we have two versions, we have used the original hand-written letters, but noted that a version of this letter was also printed as a newspaper article as well as the dates it was printed in *Vendsyssel Tidende*.

The photos in the book mainly consist of photos CL had taken and sent home. CL sometimes wrote a short explanation on the back of the photo. CL's own explanations are rendered in single quotation marks.

Measurements, currencies, etc.

Fod: A Danish fod [foot] equalled 0.31385 meters, so it is slightly longer than one English foot

Tomme: A Danish tomme [inch] equalled 2.62 cm (12 Danish inches made up one Danish foot). An English inch equals 2.54 cm

Alen: A Danish alen [yard] equalled 0.627 metres (or 2 feet). An English yard equals 0.914 metres, so it is longer than the old Danish yard.

Pund: A Danish pund (pound) equalled 500 g, so it is a little heavier than the English pound which equals 454 g.

Because feet, inches, yards and pounds are app. the same whether we use old Danish measures or English measures we have in the text used the Danish word and in brackets translated in the following manner: 3 tommer [inches].

Mil: (mile) is also an old Danish measure. One Danish mil equals 7532 metres (4.68 British miles). An old Danish mil is thus not at all equal to an English mile and we have therefore indicated this as follows: 1 mil [4.6 English miles].

Tønde (td.): land, old Danish measure. 1 td. land equals 5516 square metres or app. 1.3 acres.

Apart from old Danish measures CL also used metres, gallons, etc. reflecting that he also had lived in USA for about ten years before he went to German East Africa.

Reaumur: French measure, commonly used in continental Europe in the 1880s.Water boils at 80° Rm. To convert to centigrades multiply with 5/4 (17° Rm. equals 21° Celsius).

Currencies

GEA-Rupee and pesas: The currency in German East Africa was until 1904 the Rupee, divided into 64 pesas. According to letter from CL 25th December 1895 1 Rp is worth 1 kr. 10 øre.

Kroner and øre: Kroner (kr.) is the Danish currency. 100 øre equals 1 krone.

CL most often gives prices in kroner, but also in rupees or Mark. Reichsmark or just Mark was the German currency. Until 1914, as long as the gold standard existed, exchange rates were fixed. 1 Mark equalled 88.88 øre, 1000 Mark 888 kr. and so on.

Prices have been converted into today's prices when important. To do so, we have used Statistical Yearbook's price indexes (Statistics Denmark, 2004).

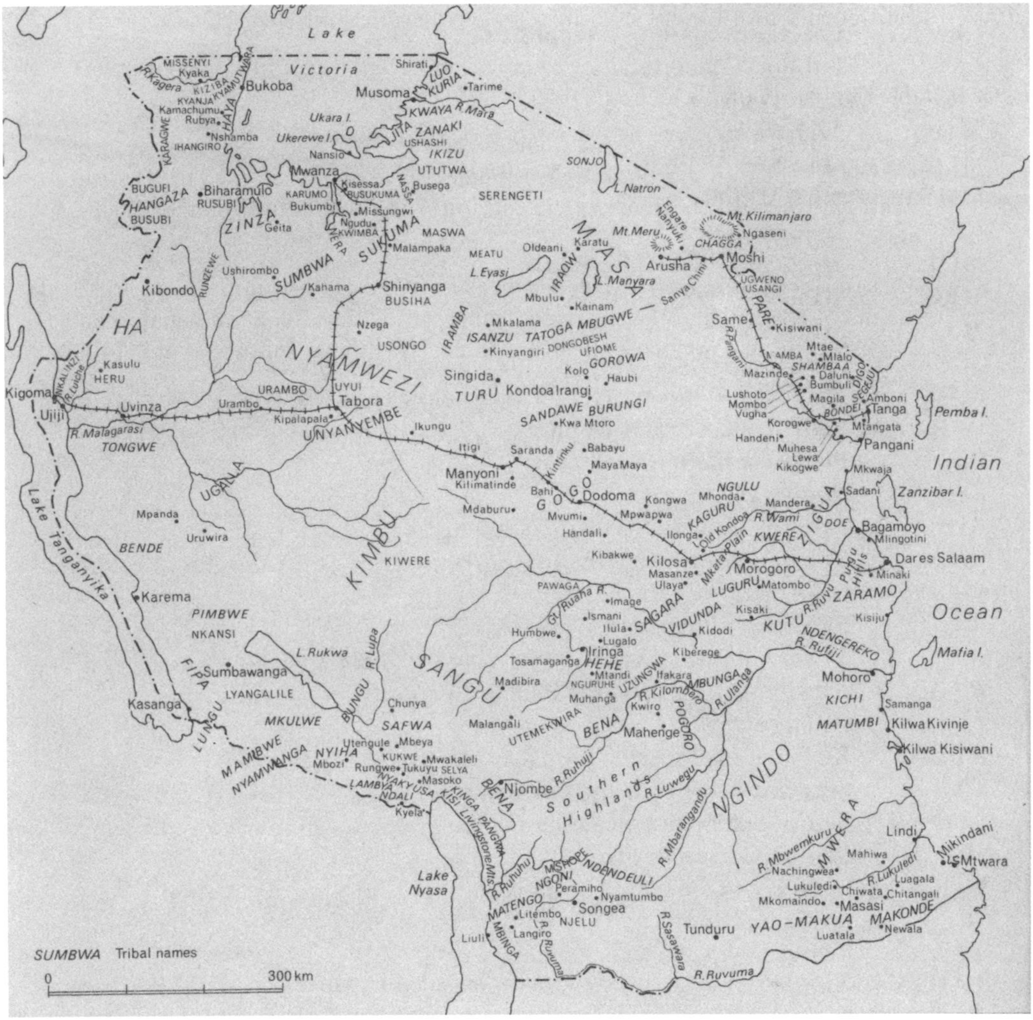

Map of Tanganyika with names of places and tribal names reprinted from John Iliffe, A Modern History of Tanganyika (1979), with kind permission

Modern map (1976) of the Northern East Coast of Tanzania

CHAPTER ONE

CHRISTIAN LAUTHERBORN: THE DANISH CONTEXT

Marianne Rostgaard

In order to fully understand and appreciate Christian Lautherborn's letters from German East Africa it is useful to clarify his background. As can be read from his letters, Lautherborn had his roots in and was formed by a distinct part of Danish society in the second half of the 1800s. Denmark was, however, not just part of his legacy. Christian Lautherborn was neither an emigrant nor a settler in East Africa; he planned from the outset to return to Denmark some day, although he never did so. Like other ex-patriates he remained a member of a community at home, although East Africa and Kikogwe (his plantation near Pangani) as the years went by, became the place where he felt most at home.

A biographical sketch: Christian Lautherborn's life and times

Christian Lautherborn was born 5th July, 1859[1] in the town of Aalborg, in Northern Jutland, at that time a rather sleepy provincial town. Christian's parents both belonged to families of non-commissioned officers. His father was trained as a type-setter. During Christian Lautherborn's childhood the family lived in a house next to the garrison hospital, on the eastern city limits.[2] The house had a garden, they kept two pigs and Christian took part in gardening and looking after the pigs as a child and, according to his sister Caroline, already then wanted to become a farmer when he grew up.[3]

[1] The Danish National Biographical Dictionary says 1860 but the parish register from Aalborg (Vor Frue Sogn) says 1859.

[2] Public census, Aalborg city, 1870, 7. Rode matr. 592. The house was owned by the garrison hospital. There were two apartments in the front house, the Lautherborn family at that time consisting of CL's father, mother, sister Caroline, aged 15, his sister Regine aged 12, CL aged 10 and Cathinka aged 8 lived in one of these apartments. The family, according to the census, had one maid. CL's father is registered as type-setter and farmer at the garrison hospital.

[3] Karl Larsen, *De, der drog hjemmefra vol. 4,* Kbh 1914. Karl Larsen spoke to Caroline Carlsen, CL's elder sister, from whom Karl Larsen also got a major part of CL's letters, the part of the letters today at The Royal Library in Copenhagen. Karl Larsen's archive in The Royal Danish Library contains notes from what must have been an interview with Caroline Carlsen, with information about the family and CL's childhood.

Christian Lautherborn's mother 1878 (Photo by H. Tønnies, Aalborg Stadsarkiv)

Christian Lautherborn's parents seem to have been upwardly mobile. His father became a compositor at the biggest local newspaper, *Aalborg Stiftstidende*—a compositor is a foreman, the person responsible for typesetting and printing the newspaper, and the family belonged, like other skilled craftsmen, to the group of respectable citizens, perhaps not too well off, but on the other hand they definitely counted themselves among and were reckoned to belong to the small town's establishment. After CL had left home the family moved to Algade 5, at the town centre, and CL's father in a letter jokes about CL's mother dressing up for an afternoon of playing whist with the wives of some of the town's well-known industrialists. This is one of the small signs, apart from the change of address, that shows CL's parents had moved up the social ladder.

Christian Lautherborn had two elder sisters, Caroline and Regine, and a younger sister Cathinka. It is Christian Lautherborn's oldest sister, Caroline, who we may thank for the preservation of Christian Lautherborn's letters. Most of Christian Lautherborn's letters from Africa were sent home to Caroline, who had married the editor of *Vendsyssel Tidende*, the local newspaper in the town of Hjørring, situated 50 kilometers north of Aalborg. Vilhelm Carlsen was in fact the son of the editor of *Aalborg Stiftstidende* so the two families, Lautherborn and Carlsen, were not only inter-married, both families were also closely connected to the two local newspapers, so the idea to have Christian Lautherborn act as a correspondent must have been rather obvious to the family.

Caroline was definitely Chistian Lautherborn's favourite sister, although his two other sisters were closer to him in age. Maybe this was because they, and Caroline's husband Vilhelm Carlsen, shared world-views. The two brothers-in-law, and Caroline, although as a woman of her time she partook in a less direct way, had a mutual interest in societal matters and were concerned with social conditions in Denmark as well as East Africa.

When he left school, probably in spring 1873, Christian was first apprenticed as a mechanic. He finished his apprenticeship, and his mechanical knowledge would prove extremely useful later, but he wanted to go into agriculture. He began employment at the manor Sohngaardsholm, within walking distance of his parent's home. In 1877 or 1878 he worked at a small manor, Vestergaard, in the parish of Skibsted, app. 30 kilometres south-east of Aalborg[4] as a junior farm-bailiff.

In November 1879 he decided to move to the USA together with one of his elder cousins. He ended up in Texas where he got work on a cotton plantation owned by a former Danish immigrant to the US. He worked as a farm-bailiff and tenant farmer on diverse locations in the Galveston area in Texas in the 1880s. He also tried owning a shop and the dairy business, to which he attached great expectations. CL did in fact earn money in Texas, but lost it again. He obviously knew more about farming than business. In 1887 he accepted an offer to manage a cotton plantation for the German East African Company in today's Tanzania, at that time German East Africa.

Christian Lautherborn in his letters home, never explained why he decided to leave Texas and go to East Africa instead, but a recurrent theme in his letters from Texas is his hope to make a modest fortune one way or the other—enough to set up himself as an independent business man, preferably in agriculture, but possibly some kind of trade as well. Being independent and self-employed seems to have been important for him. His final aim, or hope, seems to have been to return home to Denmark with money enough to buy at least a sizable farm.

Christian Lautherborn is quite typical for Danish émigrés in the 1870s and 1880s. This group was dominated by skilled craftsmen or people from the aspiring middle classes, younger sons of farmers or people who had learned a trade, in short people who on the one hand wished to keep up a position in society—and not become common workers—but on the other hand lacked the means to establish themselves as independent businessmen or shop-owners, or buy a farm at home.

[4] Trap Danmark 4th edition, Aalborg Amt, p. 666.

For one reason or the other Christian Lautherborn never returned to Denmark although he mentioned returning home from time to time in his letters. He stayed on as a manager for German East African Company from 1888 till his death in 1906 in Kikogwe, just outside Pangani, Tanzania. Again he never explicitly explained why in his letters, but reading them it becomes quite clear that during his years in East Africa he both adjusted to and became addicted to Africa and his life as a plantation manager. Some of the later letters state clearly that, while happy to be home, "There was always something I missed when I was home. It was not having anything to be responsible for" (letter, 26th January, 1901). In the end Christian Lautherborn came to lead a life as a plantation manager in Africa he could never have led at home. It was not only a question of being manager or a king in his own little kingdom, the plantation also provided him with something useful to do with his life—a theme that threads through all his letters. To make cotton, coffee and sisal production flourish in East Africa and besides improve everyday life in the region had become his life's project.

The following pages will focus on four themes highlighted in the letters where CL's background in Danish society in the late 1880's obviously had a great deal to say about his mental outlook—the way he perceived the African experience, reacted to it and acted in the African context. Class or what constitutes or justifies differences between people definitely coloured Lautherborn's way of classifying people. He had a number of values he used to judge and rank people, sometimes (but not always) in accordance with more general European prejudices of the time. The way CL constructed and justified differences between people is therefore worth inquiring into. His attitude to religion also seems worth investigating. Christian Lautherborn in his letters several times returns to controversies between two revivalist movements within the Danish Church at the end of the 1800s, Grundtvigianism and Home Mission. It is not an idiosyncrasy of Lautherborn's, on the contrary this controversy deeply coloured a number of political and cultural debates and influenced the course of Danish society in the 19th and 20th centuries. Christian Lautherborn's stand on Grundtvigianism influenced his general outlook on life and is related to his and others' belief in progress and improvement. Indeed, progress and improvement are key-words to the late 19th and early 20th centuries, but Lautherborn's background in Danish society gives these key-words a slightly different value than their connotations in a British Victorian setting.

Finally we will explore the Danish image of Africa in the years 1880–1900. Denmark was never an imperial nation, not because of any moral high grounds, it was plainly much too small a state to engage in colonial adventures in the 1800s, but the European imperial mission nevertheless had its proponents. Parts of or whole letters were printed in *Vendsyssel Tidende* as 'Letters from Africa'. Printing Lautherborn's letters in a Danish newspaper contributed to the public picture of Africa, but why was

Lautherborn's brother-in-law, the editor of *Vendsyssel Tidende*, interested in African or overseas matters? And in what ways did letters add to or comment on common Danish notions about Africa?

Upstairs and downstairs in Danish society in the industrial age

The town of Aalborg had about 10.000 inhabitants in 1860. Thirty years later the number had doubled. The 1880s and 1890s were industrial boom years with new opportunities opening up and old ones closing down. Money was invested, money gained and money lost in these years. The new elite of industrial entrepreneurs in most cases belonged to or married into the old elite of merchants and wholesale dealers or owners of large workshops. Money to invest was a prerequisite, but apart from that the 1880s were years when a wealthy hosier could found a cement factory that, also thanks to good luck, became one of Aalborg's largest industrial facilities. Old notions of ranks and class were re-shuffled for high and low as well. In the old social order, a craftsman had had a relatively secure position, safeguarded by the guilds finally abolished in 1865. Hereafter a trade became much less of a guarantee of status although differences certainly existed between skilled and unskilled workers. Small workshops lived on, but to a large degree as repair shops, much less prestigious than the old craftsman's workshop. So without money to set up some kind of business, most tradesmen faced a future as a worker, especially mechanics since independent workshops more or less disappeared with the coming of industry. Another possibility very slowly opening up in Danish society in the 1880s was to get an education and enter the professions. Altogether, however, the 1870s and 1880s were years of stagnation in a provincial town like Aalborg: new opportunities only opened up in the 1890s This may also explain why emigration from Denmark reached a peak in the 1880s.[5]

Christian Lautherborn clearly saw going abroad as an opportunity for the less well to do, as his comments in a letter about Peer Scavenius, who belonged to a family of great landowners reveal (letter, 15 March 1895).

[5] Kr. Hvidt, Flugten til Amerika [Danish emigration to America], p. 212–220 and p. 237–240. Fig. 26 p. 239 shows that the number of emigrant craftsmen peaked in the 1880s. It may be seen as a reaction to industrialisation. Regarding "farmers" and "farm-hands" the data are more unclear. However, the figures show that owner-farmers were rare emigrants. The bulk of emigrants were young people (aged 20 to 24 years). Many of these young men were most probably the younger sons who would not inherit the family-farm. For many of them there was no real future in the Danish countryside. Kr. Hvidt has not systematically researched the social background of the emigrants, but he writes p. 251, that emigrants from the countryside by far outnumbered emigrants from the towns, and that many more emigrants came from the islands in Eastern Denmark than from Jutland until the 1880's. Especially cultivation of the moors had until then supplied land to new farms.

He could see no justification for a member of the landed aristocracy to go abroad, as they had land enough at home to supply a more than decent income.

Like the city, the countryside had its clear differences between the classes. Manors and manorial lords, and farms and farmers, were two worlds apart. The manorial lords represented the old aristocracy whose power was dwindling, but the class of manorial lords remained the back-bone of Danish governments until 1901 (*Systemskiftet*). The manors cultivated around 12 percent of arable land in Denmark, the rest was cultivated by farmers or peasants in the late 1800s. Starting with the great land reforms in 1788, adscription was abolished and tenant farmers were allowed to buy their farms, which the majority did over the years. In the 1850s villeinage was finally abolished, that is the remaining tenant farmers could no longer be commanded to do work on the manor they belonged to. Instead of tenant farmers the manors hired smallholders or farm hands. This made room for a new group of bailiffs and overseers to supervise daily labour on the manors. In the 1880s and 1890s the manors also started to employ agricultural trainees more systematically. These young men wanted to train for a career as a land manager or farm-bailiff working on a manor, or maybe expected to inherit a sizable farm. The manors, at least until the 1880s, were still regarded as the most progressive part of Danish agriculture, so if one wanted to learn about new methods in dairy production or grain growing the manors were a good place to learn about agricultural improvement, although farm institutes and agricultural colleges also came into existence in the late 1800s. The Folk High Schools that mushroomed after 1864 also taught practical agricultural knowledge to their students, students that were to become the next generation of farmers.

The position as a trainee was one of the few ways into farming if one were not born into it. As a trainee one could earn some money and have free lodge and board, and that certainly would have attracted some of the trainees. The agricultural trainees were normally recruited from the middle class; their fathers usually were farmers, craftsmen or perhaps village school teachers. During their training they would learn all the different tasks necessary to master management of a big farm or manor, from raising cattle and cultivation of crops to book-keeping.

There were two sides of the coin to being an agricultural trainee. On the one hand agricultural trainees were not paid very well and had to work long hours in the fields, the stable or the barn, alongside the farm hands. On the other hand some landowners were truly engaged in spreading new, modern methods of farm management and on some manors the landowner himself or his land agent would teach evening classes to the trainees. Farming, however, was basically learned the hard way, by doing.

To be a progressive in Danish agriculture in the late 1800s meant taking an interest in ways to improve yields and altogether the economic gains from agriculture. Dairy farming (butter for the British market) was one of the areas that saw many innovations in the 1870s and 1880s—it became mechanized and the importance of hygiene was highlighted. The search for new and improved methods sparked off cooperation between the practitioners and scientific researchers. Threshing machines and other technical innovations were also slowly becoming a part of Danish agriculture. The manors were clearly in the forefront. A third area that became a kind of a popular movement and definitely a sign of progressiveness was land reclamation. Draining of low lands and moors were big and costly projects where manors also were in front, but farmers might also join company and raise the necessary capital. Altogether Danish agriculture, both on manors and farms, took a huge interest in improving methods, to a certain degree competing with each other, especially in dairy agriculture. The faith in improvement as the way ahead had thus become thoroughly embedded in Danish agriculture at the end of the 1800s.

We can see how this zest for improvement also influenced the young CL. In one of his early letters home, during his years as an agricultural trainee, CL relates that in his spare time he reads books, both novels and scientific books on agriculture. Whether he was urged to do so or it was his own idea, he does not mention.

The Danish National Museum has collected a number of written memoirs from people working on or in other ways connected to life on the manors between around 1880 and onto the 1930s.[6] All of the memoirs from agricultural trainees and other people working on the manors, tell about the long working hours. In the summer they rose at half past four in the morning and worked until 6 o'clock in the evening. In the winter they may wait until 6 o'clock in the morning to get up (it is pitch dark at that time during winter in Denmark, so nothing much could be done anyway). The trainees were, as already mentioned, employed to learn how to manage a farm, but they were also used as cheap labour. The trainees, however, worked long hours doing hard physical labour first and foremost because it was natural. Any young man living in the Danish countryside in the 1800s, apart from the disabled and the few sons of manorial land-owners, whether a farm hand, a farmer or an agricultural trainee, worked what today would be considered unbelievably hard. The trainees also worked hard to get good recommendations and thus work their way up during the manorial hierarchy. The largest estates were worlds in themselves with several layers of overseers and bailiffs working under the supervision of a land agent. A land agent or a farm manager on a manor would live in a stately home, would employ several servants and live like a master himself. He would of course never become part of the aristocracy, but could certainly live on a footing above

[6] Ole Højrup, *Herregårdsliv 1–6*, [Life on the manors 1–6] Nationalmuseet, 1980–1982.

ordinary farmers. So although the agricultural trainees worked together with the farms hands, they had a different future. They differentiated themselves and were differentiated from the farm hands. Farm hands had at the end of the 1800s become a despised lowest class in Danish rural society, in general regarded as rough hands with a predilection for hard drinking and loose living. The trainees ate in separate rooms from the rooms where the farm hands ate and they dressed differently. Agricultural trainees often wore boots, not wooden clogs, and although their salary was meagre, they spent money on proper clothing—wearing a jacket, shirt and tie whenever not doing manual work.

Chr. Lautherborn may have learned at least two things during his years as an agricultural trainee, both typical for Danish agriculture in the late 1800s. The one was to work hard. Hard work was neither valued in itself nor despised, but considered natural. The other was to relate work to progress and combine the idea of individual social climbing with progress on a societal level. Whether one was a farmer, a landowner or land agent, individual or societal progress, was seen as closely related to the practical task of farming. How to handle cows or plough a field might be learned best from an old farm hand that had done the task hundreds of times. New methods in agriculture, however, were spread from the top down and progressive agriculturalists had to be able to do whatever they wanted to be changed more or less with their own hands, and thus by example teach others.

The meaning of work was, if we look at CL's letters from his later years in Africa, connected to agricultural progress, thereby of course hopefully to earn money, but not only earn money for the company, but also to do so in a way that improved living conditions, at least for anyone who worked on the plantation and did a decent job. Hard work was thus bestowed with a higher meaning—working for a better future life. He stated this most explicitly in a letter to Caroline on 27th January, 1890:

> I do not at all wish to go to a more civilised place. I fit in very well with circumstances here and am very well pleased. I think every man has a duty to do as much good as possible in life. When one lives under more civilised circumstances it takes much more talent than I have to do good. This is not the case here where I live among half-civilised. The little knowledge I have proves itself useful every day and I can accomplish something useful every day that makes me as well as others happy.

In this sense Christian Lautherborn valued work and practical abilities. The ability to work conscientiously and accomplish something was the defining criteria in his way of ranking people and it fits in with his firm belief in status being just when it was earned and not assigned. The letters contain numerous examples of this way of ranking people. It is very obvious in his dislike of the Arabs. One quote will make do here: "previously the Arabs were the highest nobility in the land, a class of people who were far above the natives and whose influence and commands they passively obeyed.

It is difficult to say what they have done or what they do, but one thing is certain, they have never worked" (letter, 15th March 1895). In contrast he praises the *Bondei* people. When CL travelled for the first time to the Handeni mountains in 1894 he came across the *Bondei* people and their well-maintained fields. "They prepare their fields with a hoe and, in an unbelievably short time, can clear a large piece of land of grass and weeds. There is no doubt that they are good farmers." (letter, 21st January 1894). Because of that they earned his respect.

Only three letters written by the young Christian Lautherborn during his years as an agricultural trainee exist today and he does not tell much about his daily life on the relatively small manor, Vestergaard. He speaks of little incidents from daily life, about a wreck with a carriage and about a fatal accident with a threshing machine. Besides the gossip about friends and relatives, he assures his sister that he likes farming. The letters, however, reveal his hopes for the future as a kind of independent business man. They are written from the point of view of owner or manager of the farm, telling about the prices 'we' got for grain this year.

Christian Lautherborn's letters from the United States document his identity as a farmer. He writes about people and customs in his new country, but he certainly also tells about the climate, the soil, the impressive corn fields and the fat cows. One of the things he clearly valued in the US was that rank played a lesser role than it did at home. In a letter to his sister and brother-in-law in Hjørring in January 1881 he states approvingly that: "Everyone is equal here. When I talk to businessmen and other wealthy people in town I am still Mr. Lautherborn, even though I am, at this time, a farm hand." As he explained in another letter home, anyone can have quite another station in life next year, everything changes so quickly in America that people cannot be placed in categories like at home. A society promising that anyone who worked hard, diligently and conscientiously would be rewarded, was a society that fit CL's aspirations and he returns to that theme in this June 1881 letter: "Mother asked in her last letter if I was never homesick. I cannot give any other answer than no. I, of course, often think of you and wish all good luck for my old home. I have, however, no reason to be homesick, because everything is going so well; and I have good prospects of becoming something; which was after all, not the case back home."

He reverts to this theme of 'becoming something' and underlining his dislike of being somebody's junior again in an August 1883 letter: "Your wish that we may see each other in the coming year is also mine, but I do not think it will happen. The necessary money for the trip I could soon save, but travelling home would be a poor use of it. I know that you are well. Even though it is also my deepest wish to see all of you again, the joy would be short-lived, since I would have to return to America. No little sister, if I cannot come to Denmark as a wealthy man, I will never return. I could never return to work for an arrogant boss, I have lived too long in Texas for that."

Later Christian Lautherborn sought to speed up fortune-making in order both to come home to his family and realise his dream of returning as a wealthy man. He had at that time earned a considerable sum, invested them in patents, lost all the money, and on top of that acquired a debt too.[7] CL never became 'something' in Texas. After his adventure into the world of business, he returned to agriculture. He managed to repay the debt and started over, once more trying to save up money to realize his dreams.

Home Mission versus Grundtvigianism

The Danish church and Danish society was, as already mentioned, deeply affected by two revivalist movements in the second half of the 1800s, Home Mission and Grundtvigianism. They were originally one movement in the1840s, generally years of social and political upheaval that spawned religious excitement as well. In the 1860s they split, becoming staunch opponents then on. There may have been theological differences, this, however, only preoccupied part of the inner circles of the two movements. To most people the split was not a question of theology, but rather of culture and daily life. It was a matter of one's outlook on life and humankind in general (whether humanity is essentially sinful or created in God's image, ie. people are a part of God and potentially good—people of course sin, but it is a question of whether one stresses the potential sinfulness or the potential good in human nature). Both believed in salvation by faith only, not by good works. In this they were in full accord with the Danish Lutheran church, which both movements remained within, but they interpreted and handled this article of faith in two very different ways.[8] For Home Mission, faith was all that counted, and keeping away from sin, or if unsuccessful, repenting your sins in hope of God's mercy was essential. It was a sin to sing (apart from solemn hymns), to dance, to drink, to play games (and especially combine it with betting), in short as some Grundtvigians mocked, it was a sin to have fun and be happy. What counted for Home Mission was life beyond the grave, not life on this earth (to a certain degree a vale of tears). Grundtvigianism was labelled, and they called them-

---

[7] Karl Larsen, *De, der tog Hjemmefra IV. En tro Tjener*. Kbh 1914, p.8.

[8] This paragraph builds on P.G.Lindhardt, an authority on Danish church history. Apart from that I also draw a little on my own experiences—I grew up in a family thoroughly influenced by Grundtvigianism, and especially when it comes to the consequences of Grundtvigianism, the mental outlook on life, I can very easily fit memoirs of my grandparents with P.G. Lindhardt's text (and CL's letters).

selves 'the happy Christians'. Faith still counted, especially being pure at heart like an innocent child. However, one did not have to behave in certain ways. Indeed, to do so was to a certain degree despised as being a hypocritical show-off, as long as one kept God in one's heart. In fact Grundtvigians likened Home Mission's condemnation of sinful ways with a belief in salvation by good works, a heresy in the Lutheran church.

The two revivalist movements did not have membership cards etc. and as they remained within the Danish church, it is hard to say how many followers they had.[9] The movements are best likened to rings in water. There was an inner circle, that held meetings, edited magazines etc. and around them, an outer circle of people who read these magazines and attended meetings. An even more removed circle of people never went to meetings etc., but anyway the two movements coloured their outlook on life.

In daily life the difference was mainly a difference between establishment of a rather closed community of the saved (Home Mission) or a more outwardly turned, inclusive community (Grundtvigianism). Both movements saw life as a gift from God, but differed on how to treat this gift. To Home Mission advocates, this gift had to be preserved and protected in order not to disgrace it. For Grundtvigian affiliates the best way to thank for this gift was to make something useful of it. In any case Grundtvigians were much more engaged in their local communities, in politics and social life. Life on earth was not a vale of tears, on the contrary it was a gift from God and it was a sin not to enjoy this gift. Home Mission people consequently often accused Grundtvigians of excessive worldliness.

Christian Lautherborn does not talk that much about religion, but he equates superstition among the Africans with Home Mission several times, as in this quote: "They [the black people in Africa] live in fear of Satan and evil spirits just like the Home Mission people in Denmark". It is not a question of mocking religiosity in general—CL definitely believed in God: "I know what Vilhelm Beck [the spokesman for Home Mission in Denmark] says is a lie. Mankind are the children of God. He is a good and loving father, is patient with our small faults; this we learned when we went to school. Now Vilhelm Beck will suddenly give us a stepfather who has less

[9] The Danish Church has not been a state church since 1849. After 1849 (the year of the Danish Constitution) the State Church was transformed to 'The Peoples Church of Denmark' (Folkekirken) and bonds between church and state were loosened, although the Danish Church, albeit no longer a State Church had a privileged position compared to other churches in Denmark. Part of the political programme of Venstre (the Left) which Grundtvigians belonged to, was that the minister should be elected by the parish-community, not appointed by the bishop. When Venstre came into power in 1901 a reform of the Danish Church was one of the first points on the agenda. Since 1903 ministers have been appointed by a council elected by the parish-community, the bishops, however, still have to approve. Apart from the right to elect their own minister, people were, also before 1901 allowed to found their own elective community (valgmenighed) and at the same time stay within The Danish People's Church.

sympathy for us than a Sahara lion for its prey. (...) If they [a group of churchgoers in Denmark who had listened to a Home Mission inspired minister] believed what was told them, that they only lived here on earth so that their souls could be tortured in hell after death, there was good reason to look dark and miserable. I will only hope for my homeland that enlightenment moves forward faster than Home Mission." (letter, 15th March 1895). CLs dislike for Home Mission as well as superstition among the Africans is located in Grundtvigianism's practical and anti-authoritarian outlook. It rejects the notion of illness as a punishment sent from some god or evil spirits, it is just illness. People should not believe in hell and eternal sin and leave such matters to somebody else's mercy—be it a punishing God or evil spirits, but take matters into their own hands. One may perhaps also see Christian Lautherborn's educational endeavours, where he tries to teach the Africans some astronomy, as an off-shoot of Grundtvigianism's interest in education, but it might as well be a product of his agricultural training which emphasized the importance of the natural sciences. His zest for practical ways of doing things may also be a product of his training as a land manager as well as his cultural baggage from Grundtvigian circles in Denmark as both valued improvement and progress. It is, however, not only the few direct hints to religion where one can discern a colouring from Grundtvigianism. First and foremost, his way of looking at work, and the many modernising projects he pursued on the plantation, may be attributed to a Grundtvigian outlook on life.

Grundtvigianism became strongly affiliated to the democratic and national movement in Denmark in the second half of the 1800s, mixing a Hegelian inspired progressive optimism, (Grundtvig had read Hegel) rather freely with the Bible. The 'promised land' became a mixture of Bible prophesies and the coming of a better life on earth, brought forward by history's progress. As part of the national revival, and to differentiate Danes from Germans, and stress their Nordic character, old Nordic mythology became very popular and was mixed in as well—Grundtvig was a master eclecticist and obviously very charismatic. Numerous people who met him have testified to that. Some people who heard his sermons claimed the Holy Spirit had descended. In contrast, some of his adversaries claimed he was unstable and spouted a lot of nonsense.

Grundtvigianism is a non-conformist movement where faith was a personal matter and an experience, and the Bible just a book, not the revealed truth. As faith was not a question of the correct belief endorsed by the authorities (the state church represented by its bishops), everybody had to be educated, and here Grundtvigianism came to play a vital role as a progressive and egalitarian movement. It put a lot of effort into building schools, especially the folk high schools, mainly attended by sons and daughters of farmers, where practical knowledge and an 'education of the heart' were mixed. The folk high schools were originally seen as an alternative to the formal educational system (the universities) with their 'barren knowledge' and 'book-learning'. Grundtvigians emphasized what they called 'the living word' (opposed to dead book-learning), which encouraged education as a dialogue where teachers were meant to inspire and awaken, not teach a given curriculum (the folk high schools had no curriculum and no exams).

Late in the 1800s and early 1900s some people affiliated to Grundtvigianism also joined campaigns for women's suffrage and, to a lesser extent, the temperance movement. Above all they engaged in their schools and in the cooperative movements, and as already mentioned, campaigned for a constitutional reform. The political party Venstre (the Left) whose prime cause was constitutional reform was closely connected to Grundtvigianism. *Vendsyssel Tidende*, which CL's brother-in-law Vilhelm Carlsen edited and co-owned, was politically affiliated to what may be termed the left or radical wing of Venstre (venstre-reformpartiet). Ths. Bjørnbak, for a short period co-editor of *Vendsyssel Tidende*, together with Vilhelm Carlsen, was also a well-known figure in the party. Besides, being co-editor of *Vendsyssel Tidende* Ths. Bjørnbak was also the rector at a folk high school and we know that one of Caroline and Vilhelm Carlsen's daughters, Ellen, attended folk high school (letter, 17th April 1901).

Vilhelm Carlsen was engaged in numerous causes, such as The Peace Cause (see chapter six), the temperance movement and the land reclamation movement, including the reclamation of the sandy soils of North Jutland. In his capacity as editor of the local newspaper, he was deeply involved in many aspects of local life and Caroline seems to have shared his views and also to have participated in some of the many schemes, among others the founding of the local museum in Hjørring. *Vendsyssel Tidende* had an extraordinary wide circle of readers for a small provincial paper, attributed to a large degree to Vilhelm Carlsen's lively interest in a vast number of ideas and projects that agitated Danish progressivists at the end of the 1800s and beginning of the 1900s. Vilhelm Carlsen, whom we know most about because he was a rather well-known public figure in his time, definitely belonged to what was at that time the progressivists in Danish society, and engaged in improvement of society in many and rather diverse ways. The harnessing of the natural sciences to further economic development and social progress and improvement of social conditions was definitely also part of what it meant to be a progressivist around 1900.[10]

Caroline and Vilhelm Carlsen, to whom most of CL's letters were addressed, were clearly CL's discussion-partners when it came to socio-po-

[10] Dansk Provinspresse [Danish Provincial Papers], Provins-Journalistforeningen 1900–1925 pp. 88–89 and P. Horn, Hjørring i Fortid og Nutid, udgivet i anledning af byens 700 års jubilæum som købstad, 1943 [Hjørring 700 years as privileged town] pp. 206–208.

Vilhelm and Caroline Carlsen. Relatively unusual for its time the man sits and the woman stands, normally it would be opposite. No date, most probably around 1885 (reproduced from 'Billedhæfte om Caroline og Vilhelm Carlsen', 1978, VHM)

litical and moral issues. It is clear from reading the letters that these two members of the family played a special role for CL. Consequently CL's mental outlook was not only a question of upbringing and background, but also deeply influenced by his ongoing discussion with Caroline and Vilhelm Carlsen through all his years in East Africa. In this way the mental outlook of Vilhelm and Caroline Carlsen, and the circles they belonged to, served as frame of reference for CL while he was in East Africa.

This becomes clear in one of the first letters where CL discusses an article written in *Aalborg Stiftstidende*, in this case with his father. Here CL defends what he, as part of the German imperial project, is doing in East Africa (see chapter four). Responding to the frequent reports in Danish newspapers of cruelties committed by the Germans in East Africa, CL offers his own experiences as a defence and uses the even more outrageous behaviour of the Arabs as an argument against the German cruelty (letter, 13th November 1891). Another example of his Danish frame of reference surfaces in CL's discussion with Caroline Carlsen about the labour questions. "You write about the labour question and believe that it is unjust that the small man is oppressed by the big who lives at the expense of the other. But is it not that way in all of nature? See for example the shark that swallows all the smaller fish, because it is stronger, the eagle the dove, the wolf the lamb, the lion the deer." CL goes on to argue against this naturalisation of social differences, the naturalisation of social differences is not his own way of seeing things, it is a reference to what was the general opinion of that time. He goes on: "We are, however, equipped with the ability to think, which nature has not bestowed on the others. Therefore we feel it is unjust that one should live in luxury, while the other does not have enough to satisfy his hunger. We try, as best we can, to relieve the need of the poor." (letter, 21st January 1894). In other letters CL lets his sister and family back in Denmark know that he pays his workers a decent salary—the pay and its buying power is reported home rather meticulously. Even household matters and children's upbringing are discussed between CL and Caroline Carlsen. In one of his last letters, CL offers his experience from East Africa as a guideline for how Caroline should bring up hers and Vilhelm Carlsen's many children (they had 12 living children). "I now have 900 black children. In any case they all call me father. But if I tried to take care of each one as you do I would not be able to do anything. No, the older ones who have been with me for a long time must take care of the younger. In this way I train them for work." CL believed Caroline should pass many more household chores to her oldest daughters and in this way also train them to be useful members of the family and society. To be a useful member of society and do something useful with ones life was clearly the core value in CL's universe.

It is difficult to discern what was religion and what sprang out of Grundtvigianism's popularity with free-holding farmers and consequently an egalitarian outlook on society. Anyway, we are talking not of theology but a mental outlook coloured by a number of movements and undercurrents in Danish society in the second half of the 1800s. In practice Grundtvigianism, though in essence a religious movement, mainly impacted society through the political party Venstre, the farmers' cooperatives, the folk high schools and the many movements for improvements in diverse areas of society. What connected these people, mainly farmers and smallholders in the country-side, but also townspeople and part of the Copenhagen intelligentsia, was a belief in progress and, especially for the farmers and the small-holders, a struggle for recognition as equals and citizens. At the end of the 1800s, Denmark was still governed by the landed aristocracy and a conservative elite in Copenhagen, and a vital part of the radical struggle focused on education or improvement of the people. What differentiates this belief in improvement from British, Victorian belief in improvement is that Grundtvigians concentrated not on improving others, but rather on improving oneself. Whereas the British Victorians consisted of an elite improving others, Grundtvigianism inspired a much more egalitarian (and liberal) view on improvement. They emphasized that improvement should be a non-elitist practice where we educate each other; it was not defined by certain behaviour traits or outward signs of moral standards. It was much more related to doing, where one may just as well praise God by enjoying life and doing something useful with one's life, as by going to church on Sundays.

The interest in Danish society in the late 1800s in Africa

The Danish State had no interests in Africa in the end of the 1800s. Danes, however, participated in the European imperial projects of the period as mercenaries, e.g. in King Leopold's Congo, settlers, tradesmen, adventurers and missionaries. Denmark was also represented at the Berlin-conference in 1884-85 which drew up borders and divided Africa among the European powers. The conference also discussed missions in Africa, which is the main reason Denmark participated, with a delegation of representatives from *Det Danske Missionsselskab* (Danish Mission Society).[11]

Danes also participated in projects like the Freeland-project, designed to found an ideal society based on utopian socialist principles with no private property. Peer Scavenius (mentioned in Chr. Lautherborn's 15th March 1895 letter) took part in the project, which had a very short life.[12] Danish newspapers and magazines mainly took an interest in adventures and experiences of Danes abroad, although the fate of famous foreign adventurers, missionaries etc. were also reported, such as Stanley's expedition to find Livingstone. The Danish public could follow Livingstone's and especially Stanley's expedition in 1879 in Danish newspapers, and Stanley's widely published expedition fuelled widespread interest in African matters just as in other European countries. The printing of Christian Lautherborn's letters from

[11] Holger Bernt Hansen, Danskere og danskhed i Africa [Danes and Danishness in Africa], *(Københavns Universitets) Almanak, Skrive og Rejsekalender,* Kbh 1996 pp 192.

[12] Holger Bernt Hansen, *op.cit.* Peer Scavenius later wrote the book, *Frilandsekspeditionen. Dens tilblivelse, Forløb og Undergang* [The Free-land expedition. Its Origin, Course and Ruin] Kbh. 1897. The expedition reached Lamu, but progressed no further mainly due to internal quarrels between the members of the expedition.

Africa definitely was part of this growing interest in African matters in the 1880s, which was sustained by the European colonisation of Africa. The German colonial project, the coastal uprising in 1888 and other news from Africa was certainly reported in Danish papers, like any other news from abroad.

Danish writing on Africa in the 1880s and 1890s mirrors the rather diverse interests of the different groups involved in Africa. Adventurers, explorers and missionaries normally wrote in modes already existing in the 1880s and 1890s.[13] Books and articles written by adventurers emphasized Africa as a very dangerous place—more or less every day the adventurer only narrowly escapes some kind of danger. In contrast, the explorers were deeply influenced by rationalist views and economic interests. Not necessarily narrowly defined as private interest, the common good also provided a driving motive behind mapping Africa's economic potential, raw materials, useful crops etc.—what we can expect to get out of Africa or what Africa may contribute to mankind's common good. From a commercial point of view, the question was how Africa could be incorporated into an international division of labour. This imperial view had part of its origin in a rationalistic and utilitarian tradition and thus dates back to the Enlightenment. These descriptions also often discussed the 'natives' and described them, more or less strictly with a view to what they may contribute to economic prosperity (or not). African laziness was often remarked upon, and became a common stereotype. Images of Africans as uncivilized, ignorant and backward, also circulated in the writings of missionaries and, if we move beyond the Danish context, colonial officials as well. However, some Danish literature in the 1800s and 1900s also framed Africa in a romantic mode, stressing the unspoiled and the natural, seeing the Africans as nature's unspoiled children, living a happy life free from the modern world's materialism in a Rosseau-like idyll.

CL's letters reprinted in *Vendsyssel Tidende* were intended for a wider audience and reflect CL's assumptions about his audiences' expectations. He writes about the more adventurous and picturesque experiences in the printed letters. The printed letters start with his journey to Africa, which his brother-in-law had requested. The first letters from Africa (chapter three and four), telling about the war, are written in the mode of the adventurer—they tell a story about the dangers in unknown Africa. If we compare these first letters with some of the later letters when CL travels to the Handeni mountains, it is remarkable how all the dangers, like snakes or scorpions

[13] This is not a very well researched subject. When it comes to Danish travel accounts only one short article has been written: Edit Uggerhal, Danskernes oplevelse af det sorte Afrika: et historisk strejftog genem rejselitteraturen [The Danes' experience of Africa: a historical incursion into travel accounts], *Noter* nr. 109, September 1991.

and even the mosquitoes biting, seems to have disappeared from the African nature.

The most common mode adopted in the letters, however, is that of the geographer or imperial explorer, noting the different crops, the climate and locations where plantations may be erected. But CL does demonstrate a true interest in the people he comes across, noting how others do things. For example, he describes the Zulu soldiers, their adaptability and their habit of carrying little items by placing them in or behind their ears. Here CL is simply curious about how people do things—he does not judge their behaviour, it is more a kind of admiration and an interest in how others do things. This of course again could come in handy, but was definitely not his primary motivation

He never adopts a romantic mode. The only time CL comes close to this, and this is consistent through all the years, is his depiction of women, particularly attractive women. Women are constantly depicted as secretive beauties—whether young girls in Naples or African women.

Apart from the, in one sense, rather straightforward descriptions of how Africa could become useful, which might serve as a kind of guide for people who wanted to go there, descriptions of Africa in the Danish public were mainly written for people who would never even dream of travelling to Africa. Their fascination was mainly with Africa as a kind of mirror or yardstick for comparisons with Denmark/Europe. In this context Africa was either seen as a backward, dark and dangerous continent that could serve to give a favourable picture of western civilization/Europe, or the opposite, where Africa came to play the role as an alternative to over-civilized Europe. Africa became the unspoiled continent, and Africans unspoiled children (or the noble savages). These stereotypes, Africans as lazy, sensual (potentially dangerous) people or happy, unspoiled children certainly circulated also in the Danish press in the 1880s.

In his first letter from Texas, Christian Lautherborn most probably reflects what he has been told about the lazy Negroes and treacherous Mexicans (who might stab you in the back if they got a chance to do so). The only other incident he relates from Texas is some years later, when he had joined the temperance movement. He gave a speech in a Black community, which he saw as a group he would like to educate and awaken. He is, however, in this letter mainly preoccupied with the impression he thinks he made on the Negroes, not the other way round.

Perhaps influenced by experiences from Texas, Christian Lautherborn is well aware from the day of his arrival in German East Africa of differences between the Africans. They are never just lumped together as 'Blacks' or 'Negroes' but considered individuals (or at least different groups). He quickly started to teach himself Swahili, like he had done with German and English. CL is fully convinced that he represents civilisation and never questions his own values and judgement in this respect. The keyword or core of these values, and judgements made according to them, is based on his belief in the importance of doing something useful and, people's ability to do a proper job was, as already mentioned, the criteria he used to judge other people by. CL's letter from 24th July1889, one of his first letters from East Africa, reveals this way of categorizing and judging people, and also how this made him act in a specific situation. He describes an incident with some Somalis he has reproached for arriving late for work. The Somalis are insulted because they, according to CL, reckon themselves to be above other Negroes. The Somalis claim to belong to a great tribe, they are not slaves like the other Negroes. "I answered that if he wanted to be better treated than the others, he would have to earn it. I did not care that he belonged to a great tribe. When he did not do his job, he would have to accept being treated like the other Negroes." As a consequence, according to CL, the Somalis hereafter came "every day, on time to their work, since they did not want the other Negroes to believe they were treated the same way as them." On the other hand Christian Lautherborn could also distance himself from his own outlook as in this slightly self-ironic rendering of an imagined conversation. "Why should he [the Negroe] work? His house doesn't cost him anything and he works maybe an hour every third or fourth day. The rest of the time is used to eat and sleep, and to live more satisfied and happy than any millionaire. He cannot understand that others will have more, when it means work. Rather less to eat and little work than more and more work is his motto." (letter, 24th July 1889).

Christian Lautherborn here at one and the same time describes the Africans as lazy, but also satisfied and happy. What is remarkable in the quote is the distance CL maintains from European stereotypes. He is not just telling about lazy Negroes, he makes it perfectly clear that there is a rationale behind the 'laziness'. He does not question his own values, to work is necessary for progress, but he is able at the same time to render a slightly ironic picture of the 'progressive' European notions of work while still sticking to his own firm belief in progress and a good days work as the blessing of life.

It is notable if one reads through the letters from CL's eighteen years in Africa, that his own voice, based on his firm belief in doing something useful with his life as a plantation manager in German East Africa became stronger as time went by. The standard-stereotypes of Africa as a dangerous place disappear and his ability to see the Africans and describe them as they see themselves, not only seen through the eyes of the imperial explorer becomes more visible. Doing something useful was for CL a question about finding practical ways to cultivate new crops in the African soils and thus contribute to the civilisation of Africa. Africans were, like the Europeans and all other people he came across, rated according to their ability to contribute to this project. Those who contributed to further civilisation as he saw it, were people who like himself were rational and progressive, who worked to the best of their abilities and found practical ways to improve material life. They ranked high on his scale no matter whom they were, like in this remark about the Sudanese: 'The Sudanese are the bravest blacks I have known in my life. And I have never noticed that they were superstitious' (letter, 25th June 1905). His curiosity about new ways people could do things and the value he placed on the ability to accomplish something, made him more open also towards African ways of doing things and even to the possibility of combining local ways with Western practices, such as using coastal building techniques but with European design and layout.

Although he basically writes in the mode of the imperial explorer his voice differs from others because he saw the Africans as active subjects not just objects for the European's imperial projects. 'The practical imperialist' may thus sum up CL's mode or voice. This was the point of view from which he saw life, a point of view that guided his work in Bagamoyo as well as his years managing cotton plantations in German East Africa. Although he was able to reflect, even critically, on his own life, he never faltered or doubted this outlook on life in his practice. And as in the incident with the Somalis, his keen perceptiveness as an observer and his awareness of cultural differences made it possible for him to act and make people react in ways he wanted them to do, without changing either his or their general outlook on life.

CHAPTER TWO

THE AFRICAN CONTEXT: GERMAN EAST AFRICA 1888-1906

Jane L. Parpart

Introduction

Christian Lautherborn's career as a plantation manager for the German East Africa Company (DOAG) spanned eighteen years (1888-1906)—the very time when plantation agriculture was widely regarded as the best way to develop the new colony. Indeed, the introduction of modern agriculture through European run plantations was a prime rationale for colonial expansion in German East Africa[1] and elsewhere. Lautherborn took this mission seriously; it was in fact the fundamental basis for his sense of purpose and self-esteem. The letters he wrote home to his family, sister and brother-in-law, and for the audience of *Vendsyssel Tidende*, thus offer a glimpse into the life, thoughts and experiences of a practical imperialist, determined to modernize agricultural production and bring 'civilization' to the region.

Yet this seemingly simple goal soon foundered on the realities of a complex and often contentious situation. A coastal rebellion (1888-1890) broke out within months of his arrival, forcing him and the other Europeans out of Pangani. During the rebellion Lautherborn moved to Bagamoyo, one of two remaining DOAG strongholds, took up arms, discovered new talents as a builder and developed labour and managerial practices that would shape the rest of his time in the region. Indeed, when he returned to Kikogwe, his plantation south of Pangani, he brought much of his mainly Nyamwezi workforce with him. He set about trying once again to establish a prosperous, modern plantation, one that turned out a profit and provided a healthy, well-functioning environment for himself and his workforce and their families. This was never an easy task; many crops adapted poorly to plantation production, labour remained hard to attract and hold and commodity prices varied dramatically. Lautherborn succeeded more than most. He proved sisal's commercial potential, retained labour in a period of labour scarcity and won widespread applause for his well-run plantations. Yet, at the end of his life, widespread rebellion, labour problems, government reservations

[1] German East Africa is on the East coast of Africa. It became Tanganyika in 1919 when the British took over after WWI, and later merged with Zanzibar to become the Republic of Tanzania at independence.

towards the plantation 'solution' and the plague threatened even Lautherborn's well-run life.

The letters provide glimpses into Lautherborn's struggles and triumphs. Like most letters, they are fragments, revealing some things and hiding or ignoring others. In order to understand both the letters and Lautherborn's life in German East Africa, it is necessary to place them squarely in their African context. While much of the primary material for the period has either been lost or is in German, the secondary scholarship in English has grown dramatically in the last fifteen years. This scholarship, along with some classic texts, has greatly assisted our efforts to reconstruct the background to Christian Lautherborn's experiences (and letters) in GEA between 1888 and 1906.

Setting the stage: European struggles over Africa

The Germans arrived in East Africa in the mid-1800s, drawn there by the possibilities of profiting from the rich trade in ivory, slaves and tropical goods such as sugar, rice, tobacco and spices, largely controlled by the Omani Arab rulers of Zanzibar and their agents on the coast.[2] Europeans and North Americans competing for influence and trade in the region were generally based in Zanzibar and content to work with the Omani Arab ruling class. The Hanseatic League companies from Hamburg, Germany, vied with other merchants for trading privileges with the Sultan and the Arab and Indian business classes. They were based in Zanzibar and more interested in making a profit than importing cheap but poorly made German goods. The British were particularly well placed with the Sultan and increasingly throughout the 19th century, came to see Zanzibar and the coastal region as their own sphere of influence. By the 1880s, the British Consul, John Kirk, had become a powerful advisor to the Zanzibar Sultan, Barghash bin Said, and the sultan's military commander, Lloyd Mathews, was a British citizen.[3]

European explorers, missionaries and traders also competed for influence on the mainland and in Zanzibar as well. The Germans joined this struggle in the latter part of the century, led by a vocal (although relatively small) group of adventurers. Carl Peters, a rather unstable, arrogant and punitive

---

[2] Juhani Koponen, *Development for Exploitation: German Colonial Policies in Mainland Tanzania, 1884-1914* (Helsinki, 1994), 167.

[3] The two main Hanseatic merchant companies were O'Swald and Hansing. Koponen, *Exploitation*, 75, 146; Kirk was definitely the Sultan's most important advisor. Norman R. Bennett, *Arab versus European: Diplomacy and War in Nineteenth Century East Central Africa* (New York, 1986); see also Robert N. Lyne, *An Apostle of Empire: the Life of Sir Lloyd W. Mathews* (London, 1936).

German adventurer, led this move with his explorations and treaty making on the East African mainland between 1884 and 1886.[4] Determined to establish a German presence in the region, Peters and his supporters pressured the government to back their land claims (most gained through deception and fraud). They also urged the government to set up a German East Africa company, similar to the British East Africa Company and other companies spearheading European imperial ambitions around the world.[5]

While wary of expensive colonial adventures, and partly to pacify an opposition at home, the German chancellor, Otto von Bismarck, did not want Germany left behind as the struggle over European domination in Africa intensified in the 1880s. Hence, he offered to host the Berlin Conference in 1884-85, which essentially partitioned the continent for European domination, most notably for the French and English but including Spain, Italy, Portugal and Germany. While England gained control over what was to become Uganda and Kenya and a sphere of influence in Zanzibar, the Germans used Carl Peter's land claims to assert their authority over German East Africa. In hopes of ruling this new territory at minimal expense, the German government authorised Peter's proposal for a German East Africa Company (DOAG) in 1885. The new company was given authority to build railroads, access mineral rights, set up banks and print money, to administer the German sphere of influence, collect customs and set up plantations on their many land claims.[6]

In its early years, DOAG existed more as a symbol of the hope for a modernized settler and corporate dominated German East Africa than a reality on the ground. In 1886, the company had about thirty employees, many of them sick and most dispirited.[7] Two years later, DOAG had only 56 men in the region, concentrated in a few strongholds on the coast and its vicinity.[8] Nevertheless, DOAG management was determined to introduce mechanised, modern European agriculture (particularly large-scale plantations) to the region. The German colonial state supported this ambition, especially the plan to produce the cotton required by the German textile

---

[4] Carl Peters was "unscrupulous, paranoid, filled with delusions of grandeur and given to acts of cruelty." Lora Wildenthal, *German Women for Empire, 1884-1945* (Durham, 2001), 17.

[5] Jonathan Glassman, *Feasts and Riot: Revelry,Rebellion, and Popular Consciousness on the Swahili Coast, 1856-1888* (Portsmouth, N.H., 1995), 180-181. Many colonial advocates came from the German middle class. They were the backbone of the Society for German Colonization. Carl Peters, for example, was the son of a Lutheran pastor and had a doctorate in history. S. von Sicard, *The Lutheran Church on the Coast of Tanzania 1887-1914* (Uppsala, 1970), 28-29.

[6] Koponen, *Exploitation*, 55-85.

[7] Bennett, *Arab versus European*, 139-140.

[8] Koponen, *Exploitation*, 78.

industry, which was struggling with the loss of American cotton during their civil war and the increasingly competitive international textile industry. The textile industry supported a large number of workers in Germany and its survival was of intense interest to the government, despite its reluctance to get directly involved in the colony.[9]

Consequently, with very little knowledge of the region and even less sensitivity to the complexities of local politics, DOAG set about hiring various 'experts' to set up plantations and begin the transfer of power to Germans on the coast. Christian Lautherborn was hired as part of this process. He arrived in Pangani in May 1888, ready to try out his Texas cotton growing experience in the new German colony. He threw himself into cotton planting and soon had a crop about to harvest. To his surprise, an armed rebellion broke out, threatening his crops and even his life.

The African context before German rule

In order to understand DOAG's failure to predict this turn of events, we need to examine the complex world of the East African coastal region before the arrival of German 'rule'. The coast had been in contact with the outside world for the best part of a millenium. The coastal towns had long traded with Asia and the Middle East as well as enduring Portugese military occupation during the 17th and 18th centuries. The string of coastal towns stretching from southern Somalia to just south of the Tanzania-Mozambique border, as well as the islands of Zanzibar, Pemba and Mafia, were linked by a common culture and language, well-developed commercial networks and sporadic political alliances and rivalries. The shared culture was literate and Muslim, based on the Swahili language, and heavily urban biased. The economic prosperity of these coastal cities was based largely on trade in slaves and ivory, particularly after the 1700s. While the foreign demand for slaves began to wane in the early 19th century with the (at least formal) abolition of slavery, the value of ivory grew dramatically along with demands for other locally produced goods, many produced by slave labour on locally owned plantations.[10]

The Omani Arabs dominated the coastal trade boom after the 1600s, when the coastal elite invited them to help rid the coast of the Portuguese, who were disrupting the lucrative trade system. Coastal resistance to Omani domination forced them to settle on the nearby island of Zanzibar where they gradually built up their trading empire. The al-Busaidi ruling family of Oman gradually redirected the bulk of East African trade through the

---

[9] Thaddeus Sunseri, *Vilimani: Labour Migration and Rural Change in Early Colonial Tanzania* (Portsmouth, N.H., 2002), 1-2.

[10] Glassman, *Feasts and Riot*.

island. The Omani military, commercial and diplomatic strength enabled them to claim customs duties from the coastal towns and to become the main conduit for this trade. This commercial centralization intensified with the move in the 1830s of the reigning sultan of Oman, Seyyid Said bin Sultan al-Busaidi, to Zanzibar, which became the centre of world trade in slaves and ivory, with a trade network that stretched halfway around the globe.[11]

The Omani rulers sought to maintain their authority over coastal elites and the slave and ivory trade through the appointment of local representatives in the coastal towns. Until the 1870s, this control was relatively loose and informal, wherein the sultan's governors or *maliwali* (sing. *liwali*) tried to promote good relations with prominent local African families by respecting their customary privileges and providing generous patronage. The extension of Omani commercial control focussed particularly on the coastal region (the *Mrima*) directly opposite Zanzibar. The Omani state exerted its authority through its *maliwali*, military garrisons and Hindu customs officials as well as Indian traders.[12]

While the Omani-dominated Arab and Indian elite played a central role in the coastal towns of this period, the local Swahili African elite (often called Shirazi) were crucial participants in coastal economic, political and social life as well. These Shirazi families benefited from the commercial boom under Omani rule, particularly from the caravan trade and their many sugar plantations. Elite Shirazi families often adopted Arab names and lifestyles, inventing genealogies that connected them to the Middle East and Arab/Omani power. Yet this Shirazi identity was deeply influenced by the local Swahili culture of the towns and the region and porous enough to encourage ambitious locals to strive for acceptance into it. As Glassman demonstrates, peasants, porters and slaves in the coastal towns and regions also saw the coastal towns as a place where they might seek their fortunes. Caravan traders, dominated by the Nyamwezi[13] also spent time in town, often working for money while waiting for the next caravan to arrive or leave. The ambitious and the lucky were able to move up the social and economic ladder, even entering the Shirazi elite, but even the poorest hoped for material advantages from their sojourn in town. The poor longed for more opportunities and blamed their bad luck on Omani misrule. Women seek-

---

[11] Laura Fair, *Pastimes and Politics. Culture, Community and Identity in Post-abolition Urban Zanzibar, 1890-1945* (Athens, Ohio, 2001).

[12] In 1887 over 1200 Indians lived in GEA. The leaders were tightly linked with the Omani ruling elite, but Indian traders engaged in trade at all levels, from large financing to the tiny but very cost-efficient *dukas* in the rural areas. Koponen, *Exploitation*, 150-152.

[13] The area around Tabora acquired the name of Unyamwezi and its inhabitants were called Nyamwezi. They were key players in the caravan trade but also developed plantations based on slave labour. Jan-Georg Deutsch, "Slavery under German Colonial Rule in East Africa, c. 1860-1914. PhD Thesis, Humboldt University (Berlin, 2000), 15-16.

ing escape from oppressive marriages as well as economic opportunities also swelled the town populations. Many slaves also managed to earn the money to buy their freedom, or escaped their masters, adding another layer to town life.[14]

The booming caravan trade increased the wealth and power of African peoples in the interior, particularly the Nyamwezi. The caravan routes increasingly tied the interior to the coastal towns, particularly the cities of Pangani, Bagamoyo and Kilwa, each of which were centres for the three major caravan routes. The Northern route connected Pangani, along the Pangani river basin, as far inland as Kilimanjaro and the Masaii country. The central route through Tabora and over to Lake Tanganyika, connected with the Bagamoyo and Dar es Salaam region, while the southern routes ended in Kilwa.[15] The growth of trade strengthened the power and influence of African rulers on the mainland, many of who even received tribute from the coastal towns.[16] The African leaders (*majumbe)* cooperated with the Omani representatives in the coastal towns, but were never bound to them. As the European power over the Omanis was seen to increase, Omani claims on the loyalty of African leaders in the interior weakened, and the possibility of new power arrangements became increasingly enticing.[17]

Pangani, where Christian Lautherborn would spend most of his time in German East Africa, was a quintessential Swahili coastal town. The main town was situated on the north bank of the Pangani River, on the ancient river delta. The oldest settlements were on better soils several kilometres to the north, atop the high bluffs marking the edge of the river's flood plain. The town on the south bank was called Bweni, fairly near Kikogwe, which was on the south side of the river, near the Indian Ocean.[18] By the 1880s, the town was highly stratified. The main settlement on the north bank included about 200 stone houses, mostly built after 1860 for Arab settlers from Zanzibar and Oman who set up plantations, as well as Indian merchants who financed the expanding caravan trade. The stone mansions built by wealthy outsiders gave the town a physical resemblance to Zanzibar. But Pangani also had narrow streets and alleys crowded with modest thatched palm (*makuti*) roof houses where most people lived while working in agriculture, fishing, crafts and petty trade.[19]

---

[14] Glassman, *Feasts and Riots*; Deutsch, "Slavery."

[15] Stephen J. Rockel, "'A Nation of Porters': the Nyamwezi and the Labour Market in Nineteenth-Century Tanzania," *Journal of African History* 41, 2 (2000), 173-196.

[16] John Illife, *A Modern History of Tanganyika* (Cambridge, 1979), 44.

[17] Glassman, *Feasts and Riot.*

[18] There was a regular ferry between Bweni and Pangani. Naval Intelligence, *A Handbook of German East Africa* (New York, 1920), 124.

[19] Glassman, *Feasts and Riot*, 38-40. In 1912 there were 140 stone houses and 700 half-stone and half-clay houses and a large number of 'native' huts. The population totalled 3200. Naval Intelligence, *Handbook*, 172.

The town economy was increasingly dominated by slave plantation production of sugar and coconuts. From the 1860s, plantation production intensified leading to increasingly bitter conflicts between masters and slaves. However, most townsfolk were able to obtain small pieces of land for producing food. While much of this food was used for family consumption, most locals also sold food and other goods to the burgeoning caravan trade. Peasants from the region came into the towns to sell their goods along with caravan workers and slaves. Thus by the 1870s, Pangani and other coastal towns were increasingly part of an international trade network that stretched from the interior of the region to Zanzibar and its foreign trading community. The power of local leaders in the interior also grew along with the ambitions of those involved in the coastal trade. The coastal religion and Swahili culture spread into the interior as well, with many small potentates taking on Arab names, adopting Islam and town lifestyles.[20]

Thus, in the late 19th century East Africa was a complex arena of power with many players. Local chiefs in the interior vied with town leaders for wealth and political dominance. The Omani Arabs and Indian merchants grew in wealth and power. The Europeans entering this scene were hardly the most important players, particularly the Germans. The British dominated the scene with their powerful consuls and their navy squadrons, which had been engaged in suppressing the slave trade since mid-century. Between 1870 and 1888, Barghash bin Said, the most powerful sultan in Zanzibar's history, managed to tighten Omani control over the coastal towns. Omani power seemed unassailable, and Omani allies, such as Kirk and Mathews, were well known on the mainland, where they were given respect as representatives of one of the new 'players' on the East African scene. The Germans, despite their commercial presence on Zanzibar since the mid-1800s, were hardly taken seriously. Their goods rarely reached the mainland, and those that did were of inferior quality, reinforcing Germany's reputation as a minor player in European commerce and industry.[21]

German colonial ambitions in the region seemed a simple business—gain Omani and British support and domination of the coast would follow. Rule through the German East Africa Company with support from the Omani representatives on the coast presented what looked like a low cost solution, with relatively few requirements for military and bureaucratic personnel. This strategy, however, overestimated Omani power on the coast, which had been waning as British influence intensified. It also underestimated the growing disaffection and the intrigues brewing in the coastal towns and interior as ambitious players from all walks of life sought to challenge the

[20] Glassman, *Feasts and Riot*, 80-81, 102-03.

[21] *Ibid.*, 177-200.

authority of the sultan's representatives.[22] Thus, when DOAG finally managed to obtain Sultan Khalifa bin Said's reluctant approval to let the company administer the coast,[23] and moved a small number of personnel to run the coastal towns, they were unsuspectingly entering a volatile and dangerous situation

The coastal uprising 1888-1890

In mid-1888, DOAG officials arrived on the coast to take over 'their' towns. They expected the letters of support from Zanzibar would quell any local resistance, but they were soon to learn otherwise. Local political struggles over power were exacerbated by insensitive and arbitrary behaviour by the DOAG authorities, most of who were blissfully ignorant of the turbulent changes that had been taking place on the coast and the growing hostility to Omani rule. The Germans focussed their attention on the Omani representatives, ignoring and underestimating the role of many other local elites as well as the slaves, proletarians and upcountry folk crowding into the cities in search of fame and fortune.[24] This misunderstanding would have profound consequences for Christian Lautherborn, who soon found himself in the middle of an armed rebellion.

In Pangani, the company leader, Emil von Zelewski, who arrived on 6th August, arbitrarily took over some of the best stone houses and publicly humiliated both the Arab and Shirazi elites. Zelewski undermined the authority of the Omani representative (*liwali*), Abdulgawi bin Abdallah. Backed by German marines from a visiting German warship, *Moewe*, he insisted on raising both the company and the Omani flags and claimed sole authority over the flagpoles—previously solely the prerogative of the *liwali* and a key symbol of Omani authority.[25] He announced that the *liwali* was an employee of DOAG and must report to the Germans four times a day. The *Moewe* left on 18th August, to be replaced by a German cruiser, the *Carola*. Determined to bring the *liwali* to heel, Zelewski invaded the main mosque with marines from the ship, disrupting the service and polluting

---

[22] *Ibid.*, 177, 184.

[23] Barghash had agreed to this in 1887, but died in March 1888. Bennett, *Arab versus European*, 142-43.

[24] Glassman, *Feasts and Riot*. The Germans chose to rely on members of the Muslim commercial stratum and to undermine chiefly power. Steve Feierman, *Peasant Intellectuals: anthropology and history in Tanzania* (Madison, WI, 1990), 121-22.

[25] Sultan Barghash had sent emissaries out to warn local rulers against selling land to the Germans and had distributed flags to buttress the authority of his representatives. Thus DOAG efforts to take over control of the flag and raise their own as well was seen as a challenge to Omani authority. There were incidents over this in Tanga and Bagamoyo as well as Pangani. Bennett, *Arab and European*, 146-47.

the mosque with infidels in shoes and a dog. He also opened the jails and let prisoners out into the streets. These actions undermined the already delicate balance of power in the town, weakening Omani authority and encouraging local crowds and ambitious leaders to challenge Omani power, and the Germans, who were increasingly seen as Omani clients.[26]

On 23rd August 1888, the *Carola* left. The Shirazi patricians and Arab elite left the town for their country estates, and crowds poured into the towns for important festival celebrations. Zelewski, who had only hoped to chastise the Arab elite, and still hoped to rule through the channels of Omani authority, continued issuing decrees and proclamations which were generally ignored by the public and the local elite, who regarded their Omani overlords and the Germans with increasing resentment. Indeed, the Omani rulers had lost most of their credibility through their inability to defend the town and its religious practices from the Germans, particularly important during the major religious holiday underway at the time.[27]

In early September large crowds swarmed into Pangani to celebrate the holidays. Tensions grew as the crowds threatened to challenge both Omani and German authority. As Lautherborn reported, a *dhow* carrying over a thousand kegs of gunpowder for upcountry caravans arrived on 3rd September. Zelewski tried to stop its distribution to the local soldiers, but failed as the crowds and soldiers were emboldened by the failing Omani authority and the small number of Europeans. Local Omani supporters, including most leading Arabs and a handful of wealthy Swahili patricians and Indian merchants, fearing the consequences of attacks on the Germans, decided to lock them in their station/custom house and set a guard of Arab mercenaries to protect them from the crowds. Lautherborn's letters describe a tense and dangerous period when the Europeans clearly feared for their lives. As matters worsened, Sultan Khalifa grew increasingly anxious about possible repercussions for any attacks on the Germans and sent his military commander, Lloyd Mathews, to rescue them and bring them to Zanzibar. On 8th September, a relieved and thoroughly frightened group of Europeans scrambled aboard Mathews' ship—Christian Lautherborn among them.[28]

Soon the Germans had been thrown out of all but two enclaves on the Tanzanian coast (Bagamoyo and Dar es Salaam). The rebellion was based on a broad but brittle alliance of Shirazi patricians, lower-class towns-people, peasants and slaves from the surrounding countryside along with some warriors and their followers from the interior. Two charismatic Shirazi elites, Bushiri bin Salim, a plantation owner from Pangani area and Bwana Heri,

[26] Glassman, *Feasts and Riot*, 202-03, 215-16.

[27] *Ibid.*, 217-218.

[28] Most of the crowd just wanted to get rid of the Germans, but five representatives of the increasingly small group of moderates accompanied the Europeans to Zanzibar to explain events from their point of view. *Ibid.*, 215-224.

a Swahili speaking plantation owner who had successfully resisted Omani rule in the Saadani region for several decades, led the fight. However, no clear lines of authority or loyalty existed. The rebels were primarily determined to destroy Omani authority, but many had local struggles in mind and were not strongly committed to Shirazi rule. Thus both the leadership and the goals of the rebellion were fluid and difficult to gauge.[29]

Lautherborn, who heartily disliked the hierarchy and political intrigues in Zanzibar, and longed to do some practical work, volunteered to move to Bagamoyo. He arrived there in late September. Like most Europeans, he thought the rebellion would soon subside and DOAG would reassert its rule.[30] Bagamoyo was a major centre of the caravan trade and many caravan personnel and local folks joined the Germans, preferring peace with trade to war. However, the apparent peace soon began to unravel. In September, an initial skirmish failed, although causing considerable damage to the infrastructure. The quiet afterwards inspired hopes that peace would prevail. These hopes were dashed when Bushiri and his retainers attacked the city in November and again in early December (1888). Bushiri remained in the region for six months, during which numerous military encounters took place and much of the city was destroyed.[31] Lautherborn provides an important eyewitness account of these struggles.

The continued attacks by the rebels, particularly the very public expulsion of Zelewski and his colleagues from Pangani, wounded German national pride. Blaming the rebellion on Arab slave traders, Bismarck managed to gain support from the humanitarian lobby that normally opposed colonial adventures. In early 1889, the Reichstag voted two million marks to suppress the uprising. Bismarck appointed Hermann von Wissmann, an experienced traveller and hard-nosed military man, as commander of the German forces. Supported by a British-German blockade of the coast, Wissmann arrived on the coast in April 1889 with a platoon of hardened soldiers from all over Africa.[32] Lautherborn shared lodgings with the German officers, and participated in some of the military expeditions. His letters describe the crowded conditions, strenuous fighting and daily dangers.

---

29 *Ibid.*; Koponen, *Exploitation*, 80.

30 Many of the Germans with DOAG "thought themselves a master race, and expected natural subordination from Africans." Jonathan Glassman, "Social Rebellion and Swahili Culture: the response to German conquest of the Northern Mrima 1888-1890," PhD University of Wisconsin (Madison, WI, 1988), 428.

31 Glassman, *Feasts and Riot*, 6-8, 210.

32 Koponen, *Exploitation*, 69, 133. W. O. Henderson, "German East Africa, 1884-1918," in V. Harlow *et al*, *History of East Africa* , vol. II (Oxford, 1965), 129. The force included about 600 Sudanese, mostly unemployed former soldiers, 100 Zulu (Shangaans) from Mozambique, about 80 East African soldiers and 40 Somali, commanded by 88 German officers and NCOs. Erick J. Mann, *Mikono ya damu: "Hands of Blood:" African mercenaries and the Politics of Conflict in German East Africa, 1888-1904* (Frankfurt, 2002), 50-53.

He was particularly impressed with the adaptability and tenacity of the Somali and Zulu soldiers, whom he regarded as clever, skilled fighters.

As Glassman has pointed out, the initial rebellion was in many ways more directed against Omani rule than the Germans. Some of the early treaties signed with the Germans were by African and Shirazi leaders who hoped the Germans would help reduce Omani influence. When the Germans claimed authority in the name of the sultan, many locals began to see them as Omani clients and turned against them on that basis. Anti-German feeling only developed after mid-1889 when the Germans returned with the full backing of the Imperial Navy and it became clear that Germany was now the colonial power to reckon with.[33]

However, once German firepower (particularly the fully automatic Maxim gun) was in place, the main towns were quickly retaken. The presence of the soldiers secured Bagamoyo from further attacks. Pangani fell in July 1889. Bushiri was betrayed and hanged in December 1889. Peace was negotiated with Bwana Heri, who continued to retain considerable autonomy in Saadani and its hinterland. Conflict continued here and there into 1890, but never with its earlier intensity. Gradually many peasants and leaders from the interior and the coast reluctantly put down their arms, but retained their determination to control their own destiny as much as possible within the constraints of the new power in the region—the Germans.[34]

Bagamoyo: the Europeans, rebuilding the city and early labour practices

Lautherborn remained in Bagamoyo until early 1891. During that time he became well acquainted with the European community, offering a picture of celebrations such as the party for Morton Stanley's rescue of Emin Pasha.[35] His comments on the personal strengths and weaknesses of the two 'heroes' are particularly interesting. He also provides marvelous accounts of the Holy Ghost Father's Order, a Roman Catholic mission that had been in Bagamoyo since 1868.[36] The Order had maintained neutrality during the

[33] Glassman, Feasts and Riot, 187.

[34] Mann, *Mikono*, chapter 2.

[35] Emin Pasha was a German expatriate working as Governor of Southern Sudan. An Islamic uprising cut him off from the world, and in 1887, Henry Morton Stanley, the renowned explorer, headed up a British rescue mission and brought Pasha to Bagamoyo. Later Emin Pasha was murdered while working for the Germans in the interior of GEA. Iliffe, *Tanganyika*, 103-06. For another eye view account of the expedition, see Jacob Wasserman, *Bula Matari: Stanley, Conqueror of a Continent* (New York, 1933).

[36] For more on the Holy Ghost Fathers, see John A. P. Kieran, "The Holy Ghost Fathers in East Africa, 1863 to 1914," PhD thesis, University of London, 1966. See also, Iliffe, *Tanganyika*, 84-5, 216-18, 232-33.

rebellion and had become a center for many refugees from the conflict. Brother Oscar, justly famous for his hospitality and his many practical talents, became a firm friend and Lautherborn's description of their times together offers rare insights into the way like-minded Europeans sought to help each other out.

While in Bagamoyo, Lautherborn discovered a talent for building. DOAG hired an architect to rebuild the war torn city, but he required assistance. Lautherborn was drafted, and soon became an accomplished builder. Indeed, he rebuilt most of the city in the two years he lived there. While predictably European in his determination to build wide, straight streets and buildings with precise corners and straight lines, Lautherborn revealed his openness and interest in local knowledge by adapting local techniques and materials in his building.[37] By 1890, he had a workforce of 60 masons, ten carpenters and 150 workmen, which built both official buildings as well as the smallest homes. In January 1890, he writes of his love for Bagamoyo and its people, and his joy strolling "along the streets here, lying beautifully even, where sun and wind can reach into every corner, knowing that I have accomplished this."

He also developed many of the labour policies that would characterize his management style at Kikogwe after his return in 1891. In particular, he demonstrated an even-handedness and willingness to teach his workers that was unusual in a colony characterized by harsh, arrogant behaviour towards African labour. Indeed, he was known "as one of those people who are of a pleasant disposition whom all the world likes." Richelman, the Station Chief at Bagamoyo, recalled that Lautherborn "had no enemy, not even among the Arabs and Negroes. With particular regards to the latter, (Lautherborn) understood how to deal with them expertly; all the Black men, women and children liked him, in whose eyes he was the "Bwana Mzuri" (the sweet or good-natured man)."[38] This approach contrasted sharply with the harsh practices of many DOAG plantation managers, which resulted in labour shortages and led to hiring slave labour and recruitment of workers from as far away as China and India. Conditions were so poor, China and India soon declared GEA off limits for their workers. Friedrich Schröder, based at Lewa plantation, not far from Kikogwe, was one of the worst

---

[37] G. Richelmann, who served on the Wissmann expeditionary force in 1889 and became acting Station Chief of Bagamoyo after May 1889, lauded Lautherborn, who managed to transform himself from a farmer into a builder, "but since he was so clever and practical, Herr Lautherborn accomplished amazing things here, and I was soon greatly in his debt." Translated by Steven Fabian, Dalhousie PhD student, from Dar es Salaam archives. G. Richelmann, *Meine Erlebnisse in der Wissmann-Truppe* (Magdeburg, 1892), 134.

[38] *Ibid.*, 134.

offenders and his plantation, not surprisingly, was constantly plagued with labour shortages and dessertions.[39]

Most of Lautherborn's employees were Nyamwezi, many of whom sought waged work while waiting for opportunities in the caravan trade. The Nyamwezi had a long, rather uneasy relationship with the Omani Arabs and their allies, who often tried to undercut Nyamwezi control over the caravan trade. Consequently, they tended to prefer the Germans to Arab/Shirazi employees.[40] Many of Lautherborn's Nyamwezi workers had worked with him during his building career in Bagamoyo. While critical of their inability to keep to straight lines, Lautherborn clearly appreciated their ability and intelligence. Indeed, when he told them he was returning to Kikogwe and could only pay some of them to come with him, his entire workforce insisted on coming, saying that "you have been our father for over two years, and, when you leave, your children will go with you ... If you won't take us with you, we will go to you; we know where you are going" (letter, 1st March 1891). And indeed, most of them sailed to Pangani with him. He repaid this loyalty in kind—arranging housing and food for his workers and their families being his first priority upon landing at Pangani.

The plantation years: production, labour and daily life

Lautherborn was hired by DOAG to plant cotton, using the skills he had developed in the Texas cotton fields. While his initial crop and most buildings were destroyed during the coastal rebellion, when he returned to Kikogwe in 1891, he quickly set about rebuilding and planting cotton. Indeed, in 1893 Kikogwe had 276 acres under cultivation of which 134 were planted with cotton.[41] He eventually managed Mwera, a nearby plantation, as well. Visitors commented on the "immaculate impression, especially the crops at Mwera which are beautiful." He continued to be an exemplary plantation manager until his death in 1906.[42]

Throughout this period (1889-1906) DOAG officials continued to believe plantations were the only way to bring modern agricultural techniques to the colony. Despite continuous problems attracting and retaining labour for the plantations, DOAG officials were obsessed with proving that the plantation option could work. Ignoring the fact that Kikogwe was almost the

---

[39] Koponen, *Exploitation*, 330-39. Glassman, *Feasts and Riot*, 187-89. Personal communication from Steven Fabian from the files of the German East African Plantation Society, Dar es Salaam.

[40] Glassman, *Feasts and Riot*, 213, 230.

[41] Sunseri, Vilimani, 13.

[42] AL2/60/39 Zanzibar National Archives. The quote is from around December 1905. Found and translated by Steven Fabian in Zanzibar.

only plantation in the country with sufficient labour, DOAG officials complained that Lautherborn was "very indulgent and good-natured towards the workers, and the performance of his people is therefore rather slight."[43] They also inspected the book keeping of their managers, driving Lautherborn to such frustration that he fired off an angry letter, arguing that he was hired as a farmer, not a bookkeeper. The company continued to invest in machinery and plantation infrastructure and Lautherborn's mechanical knowledge was clearly appreciated and respected. He was sent to America and other places to buy farm equipment, particularly the heavy machinery required by sisal production, and consulted with DOAG head office in Berlin on his trips home to discuss the machinery and production problems. Thus, during Lautherborn's tenure in GEA, DOAG remained largely committed to plantation agriculture, despite often disappointing returns and growing competition from small-scale African producers.[44]

The colonial government, which had taken over the administration of GEA from DOAG in 1890, for the most part supported the plantation option during this period as well. While some colonial governors questioned the potential of plantation agriculture and resented the demands of European settlers,[45] most government officials assumed that plantation agriculture was the key to introducing modern commercial agriculture into the region. The government was particularly concerned with cotton production, as long as cotton was in demand by the powerful German textile industry.[46] The Biological-Agricultural Institute was set up in 1902 at Amani to assist plantation agriculture, improve African production and investigate plant disease and possible new crops for the region. The Institute experimented with rubber, coffee and sisal, working closely with DOAG officials. While increasingly positive about African smallholder agriculture, until the widespread rebellion in 1905-6 (see below), government research and policy was largely framed within the assumption that plantation agriculture was the future of GEA.[47]

While DOAG plantation managers were largely independent and responsible only for their own enterprise, the company and to a lesser extent

---

43 Koponen, *Exploitation*, 350-51. This is a quote by Bericht Hindorf. By 1903, Kikogwe was experiencing some labour problems.

44 For more on DOAG, see Koponen, *Exploitation*.

45 *Ibid*., 188-196.

46 Sunseri, *Vilimani*, xxii, 17.

47 *Ibid*., 17. Koponen, *Exploitation*, 252. Lautherborn continually experimented with plants and animals as well as general labour conditions on his plantations. Wissmann came to the plantation and reported that "Herr Lautherborn fetched me ... with a cart pulled by two beautiful white donkeys, and demonstrated that the crosbreed of Muscat and East African donkeys is not only well-suited for riding, but also for working. He owns 70 work-donkeys, all of which are a result of his crossbreeding efforts." Bundesarchiv, Berlin, R1001/237a, 24/8/1895. Inspection tour of Major von Wissmann.

government officials, had some influence on crop selection. Cotton production, as we have seen, remained a preoccupation of the government and it is not surprising that Kikogwe focussed on cotton in its early years. Indeed, Lautherborn, much to the company's delight, increased his total output by encouraging African smallholder production of cotton, which he bought and cleaned at Kikogwe.[48] However, prices for cotton declined in the 1890s,[49] and plantation production continued to face labour and environmental problems, encouraging managers such as Lautherborn to experiment with other crops.

Robusta coffee grew naturally in the region, and the Holy Ghost Fathers, who introduced the arabica variety, which Europeans preferred, managed to establish a successful coffee plantation in Bagamoyo in 1877.[50] Inspired by this example, in the early 1890s DOAG set up two coffee plantations in East Usumbara, with over 200,000 trees. Lautherborn was sent there to supervise harvesting the crop because the managers were on their home break. Impressed by the quality of the harvest, he decided to experiment with coffee at Kigowe when he returned home. While plantation coffee production was initially very successful, it foundered after 1895 due to disease, falling prices and labour problems as well as questions about the soil quality. It also had trouble competing with smaller producers in the region.[51]

Rubber also grew wild in the region, particularly in southern GEA. The world demand seemed insatiable and prices were escalating dramatically. DOAG officials, hoping to benefit from the boom, encouraged their plantation managers to plant rubber. Lewa set up a large plot in 1902, and Lautherborn planted some at Mwera and Kikogwe, which did very well. Indeed, Lautherborn was so optimistic about rubber that he encouraged his nephews to join him so they could make their fortune. In 1900, rubber became the leading export product of GEA, so Lautherborn's optimism is understandable. Prices reached a peak in 1908, but sadly, the boom collapsed in 1913, reaffirming once again the difficulties facing cash crop production in the region.[52]

---

[48] Koponen, *Exploitation*, 214. Lautherborn thus foreshadowed the later shift to African smallholder cotton production. Steven Fabian, report on German East African Plantation Society files, Dar es Salaam.

[49] The GEA government did not give up on cotton, especially as cotton prices increased after 1900, but no longer saw plantations as the solution. Iliffe, *Tanganyika*, 145. Sunseri, *Vilimani*.

[50] Brad Weiss, *Sacred Trees, Bitter Harvests: globalizing coffee in Northwest Tanzania* (Portsmouth, N.H., 2003), 18, 141.

[51] The world coffee price halved between 1890 and 1898. Iliffe, Tanganyika, 127-28. Koponen, *Exploitation*, 205. H. Brode, *British and German East Africa: their economic and commercial relations* (London, 1911), 97-98.

[52] Koponen, *Exploitation*, 201-204. Iliffe, *Tanganyika*, 145-46.

The most successful transplant to GEA was sisal, a plant that flourished in the shallow laterite soils of Pangani and Tanga districts. It was indigenous to Central America, and could be used to make twine and rope, both increasingly required by the expanding world agricultural market. DOAG officials and colonial agronomists had been looking for other crops that would flourish in the GEA environment, and sisal seemed a logical candidate. In 1893 the first sisal plants were sent to GEA to be nurtured and tested as commercial crops. This experiment was the brainchild of Dr. Richard Hindorf, an agronomist who worked for DOAG. Only sixty-two plants, out of an initial 200 imported from Florida (Mexico had banned export of the plants to protect their growing sisal industry) reached GEA, where they were put in Lautherborn's care. Indeed, George Lock, a colonial agronomist, identifies Kikogwe as "the birthplace of sisal in East Africa." The decision to put the bedraggled plants in Lautherborn's care reflects his reputation as an outstanding agriculturalist. And their faith in him was well placed. By 1898 the 63,000 plants had grown to 160 cm and the first sisal fibre had been produced. Three experimental bales were sent to Germany that year. By 1900 there were over 500,000 sisal plants in GEA, mostly at Mwera/Kikogwe plantation. By 1902, this number had risen to 1,600,000. In 1900, Lautherborn sent seven and a half tons of sisal fibre to Germany for sale.[53] He became a key player in the DOAG sisal industry, making trips to America to purchase machines for processing the plants and expanding his own production so much that he had to purchase a small yacht to carry his produce to the larger steamers that could not enter the shallow harbour at Pangani.

Sisal would become the leading crop in the region. By 1902 it was the third largest commercial crop in value in the country after coffee and coconut. In 1928 it had become the principal cash crop of Tanganyika (formerly GEA). Inevitably, competitors soon emerged. In 1900 the Bushiri Estate, near Pangani, became the first company formed solely for sisal-growing. By 1905 plantations were springing up in the Tanga area as well, no doubt facilitated by the railroad from Tanga completed in 1893.[54] While he remained a leading figure in the business, Lautherborn learned the hard way that he had to guard his hard-earned knowledge about sisal production as others were ready to use it for their own gain.

Lautherborn, like other DOAG managers, had considerable control over his workforce and the daily management of his plantations. He was in charge of hundreds of workers and their families, the plantation equipment, ev-

---

[53] G. W. Lock, *Sisal: 25 Years of Sisal Research* (London, 1962), 2-3, 41. Lock was in the colonial agriculatural service in Tanganyika from 1930-1960. Koponen, *Exploitation*, 206. Iliffe, *Tanganyika*, 146-47.

[54] Lock, *Sisal*, 2-3. Koponen, *Exploitation*, 206.

eryone's health and well-being as well as the plantation buildings, the organization of his household and the shipment of goods to market. Indeed, a medical inspection tour in 1896 reported that "The labour demands [at Kikogwe] are not excessive and the treatment of the workers is good which is evident in the fact that not a single contract labourer has ever run away." Health conditions were excellent and illnesses rarely occurred. "The people look thoroughly well nourished and I found no evidence of maltreatment, nor did anyone complain of the same.[55] Lautherborn took his duties seriously and maintained a constant watch over all aspects of plantation life. He seemed happiest when he was on his plantation—indeed he wrote a letter home near the end of his life, exulting in his life, with the lovely house on a bluff, the view of the ocean, the well-oiled production system and a contented workforce. It is, he wrote, like being "a little prince in Africa" (letter, 27th August 1904).

While clearly regarding himself as the 'father' and protector of his African workers, Lautherborn's labour policy seems to have been influenced by his more egalitarian Danish background. Unlike many of his German colleagues, who were noted for the arrogant and cruel treatment of their workers, Lautherborn was seen as too easy going with his workers, and blamed for their lower productivity.[56] He quickly learned Swahili and his letters reveal many interesting conversations where he attempted to answer his workers' queries about matters such as why Europeans were so concerned about finishing work on schedule, why they wanted to grow so much cotton, the history of the Portugese in the region and other topics. Always in search of better ways to do things, Lautherborn was always interested in practical solutions to problems. When he travelled to Tanga in 1894 he noted the jars of thin baked clay the Africans stored water in and how this storing method kept the water wonderfully cold (letter, 21st January 1894), and he, as already mentioned, adapted local bulding techniques and no doubt sometimes sought advice from his workers, especially the supervisors, on African methods of farming and building. Indeed, he writes admiringly of the farming skills of the Bondei people, and of his workers' ability to size up Europeans, to give them nicknames and to mock their behaviour in painfully effective, if often cruel, pantomimes.

Lautherborn's letters support Sunseri's assertion that the plantations in GEA were never run on purely European lines.[57] In order to attract labour, the workers had to be offered conditions that would keep them on the plantation. Lautherborn clearly grew accustomed to the insertion of African

---

[55] Bundesarchiv., Berlin. R1001/126, 2/1/1897. Medical Inspection Tour of all plantations in Tanga and Pangani provinces in late 1896 by medical doctory, Dr. Gaertner.

[56] Koponen, *Exploitation*, 350-351.

[57] In 1903, 10,000 African workers were needed by DOAG and settler plantations in GEA, causing a growing labour problem. Sunseri, *Vilimani*, 54, 70.

practices into life on the plantation. The constant drumming and singing in the evenings annoyed him at first, but he soon became used to it—even missing the hum of daily life among the workers when he was in Europe. In reality, he had little choice as labour shortages plagued the region during these years, and only Lautherborn's reputation as a caring manager enabled him to maintain an adequate workforce. Even he experienced some labour shortages after 1900. In 1905, Kikogwe was only able to keep three of four sisal factories busy because of insufficient labour.[58]

The shortage of workers also meant that the DOAG plantations often had to take workers where they could find them. This led to considerable diversity in background and skills—for Europeans as well as Africans. Lautherborn experienced this most starkly at the coffee plantations in Usumbara, where he had to manage workers from around the world, including China. Eight different languages were spoken, with Swahili as the main language. However, most Chinese workers only knew their own language. Lautherborn's attitude towards his workers is reflected in a letter describing his attempts to converse with a Chinese labourer.

> There is, for example, my gardener. He is a very important person for me, since I set a high price on all kinds of vegetables and would like to give him instructions and help him with the garden. But he only understands his native language. It must be funny to see, when we two try to make ourselves understood to each other. It probably looks like two deaf-mutes, because we speak only by sign. However, we understand each other, and I think the Chinaman is happy when I come to speak to him; but I couldn't say for certain. (letter, 15th March 1895)

Lautherborn's concern for his gardener's welfare and state of mind stands in stark contrast to the cruel treatment of so many other DOAG managers. It also explains why he so rarely experienced labour problems on his plantation.

Slave labour continued to be used on the plantations in this period as well. Although officially opposed to slavery, the colonial government actually never prohibited slavery in GEA. Arab and Shirazi plantations continued to produce goods with slave labour throughout the German period. Slaves also sold their labour to plantations in order to buy their freedom, and many Germans participated in this scheme. No doubt many of the DOAG plantations had slave labourers, especially in periods of labour scarcity, or on plantations where the manager's bad reputation drove independent workers away.[59] While acknowledging gradations in treatment of slaves,

---

[58] Personal communication from Steven Fabian, from file AL2/60/39 at the Zanzibar National Archives. Written around 1905.

[59] The variety of ways that slaves managed to work on plantations reduced the difference between slave and free, making flight a viable alternative. Deutsch, "Slavery", 110, 166, 232-239, 253-274.

Lautherborn never speaks of slave labour on his plantations. One assumes that his and his family's liberal leanings would have kept him from hiring slave labour, although doing so during labour shortages must have been difficult. He does write with considerable satisfaction of outwitting a slave owner, whose concubines begged for help. Lautherborn dressed them in men's clothing, taught them how to walk like men and smuggled them out of Pangani right under their master's nose (letter, 25th December 1895).

Conclusion

Lautherborn died of black water fever (a kind of plague) in 1906, towards the end of the *maji maji* rebellion. The rebellion was a turning point in GEA colonial policy. It demonstrated African labour's power to disrupt plantation agriculture and the futility of reliance on the plantation option. It spurred a major national debate in Germany, leading to a call for reforms, including scientific development and better treatment of the indigenous peoples. In response, a strengthened colonial office abolished forced labour and introduced legally binding labour contracts. Small-scale family based agricultural production became the new 'solution' for colonial development. Of course, much continued as before. The Germans continued to rely on Arab plantations and sisal, which required heavy machinery and large-scale production continued to be grown on plantations. But plantation owners would never have as free a hand as they had before 1906.[60]

This new order would no doubt have been a more comfortable world for Christian Lautherborn, who had come to love the people of German East Africa and who remained convinced of the importance of imparting technical skills and education to Africans. While still a man of his time, with prejudices that reflected his position in the imperial enterprise, particularly against Arabs and Indians, Lautherborn's fundamental generosity and his respect for practical knowledge and hard work shaped his relations with people from all walks of life. As he wrote to his sister:

> You state in your letter that you would like to drag me out of Africa. I cannot say I wish to leave. I am fine. I have become acquainted with circumstances and speak Swahili really well. That the situation in between may be a little dangerous cannot be denied, but I do what ever is possible to protect myself. If something is going to happen I always know in advance because the negroes like me and will let me know, and I will then take appropriate action. The negroes have four names for me, *Bwana Uzuri, Mlele, Rafiki* or *Fundy Mkubwa*. The first means the handsome master, it must be ironic because I am definitely not handsome. The second name I have got from the *Nyamwezi*, it is not a Swahili name, but it means the same as the third name 'friend'. The fourth

[60] Sunseri, *Vilimani*, 2,136-37.

> means the great master or artist. From this you may get an idea about what the Blacks think about me. (letter, 27th January 1890)

It is his approach to life that may explain why he found such satisfaction in his life in German East Africa, and why so many people from all walks of life appreciated him. Thus, Christian Lautherborn reminds us that even within an exploitative imperial structure, despite its many limitations, genuine respect, sharing and empathy across apparently unbridgeable difference, could and did occur.

CHAPTER THREE

JOURNEY TO AFRICA AND SETTLING IN

Journey to Africa:

[Letter printed in VT] 2nd August 1888

From Hjørring to Zanzibar
Letter from East Africa to *Vendsyssel Tidende,* 25th May 1888

After our departure from the railway station at Hjørring, we travelled slowly by train to Hamburg. This is so well known that there is no reason for dwelling on my experiences there.

On the 4th of April, at 10 in the morning, we sailed from Bremen to Bremerhaven, where our steam ship, *Nidar*, lay rocking out in the channel, being too deep keeled to tie up to the pier. We had to be transported, therefore, to the larger ship in a little steamboat. At noon, we lifted anchor and sailed out into the North Sea. All of the passengers were German, except for a Japanese and myself. The Japanese was a relative of the King of Siam. He is in German service to learn the military profession thoroughly and will, when he has completed officer school in Germany, receive a place as a colonel in the Siamese service, from there to rise in honour and dignity. He was, however, as it was later revealed, not the only heathen present. Even though the North Sea was calmer than I had ever seen it, many of the passengers had to sacrifice to Neptune [throw up].

We arrived at Antwerp about eight in the evening the next day, and, naturally, had to go ashore to see the town. That evening we got no further than a couple of fine restaurants, but the next day we were up early and went ashore. I had thought that Antwerp was built in an old fashioned style, but that was a misapprehension, at least for the part of town I saw. With the exception of a couple of old churches or cloisters the city consisted only of brand new, wonderfully constructed buildings. One immediately sees that it is a great trading centre. For more than 1 mil [4.5 English miles], parallel to the town there is a splendid wharf of large, square stones. The entrepôt here is covered the entire length, so that people can work whatever the weather.

Parallel to the buildings there are three rows of tracks packed with railway cars, either being loaded or unloaded. There is life and activity around the clock. Hundreds and hundreds of ships lie here, and nearly all nationalities are represented. One sees Germans with their red caps, Chinese with their long pigtails, Arabs with their turbans, Japanese and Negroes. Danes and Jutlanders are seen here at the harbour every day. The town is exceptionally well built. One of the thoroughfares is at least six times as wide as Østergade in Hjørring.[1] The road was divided by a grass strip and planted with trees on both sides. Here and there, memorials for famous men are surrounded by gratings, within which are planted beautiful, ornamental bushes, and, during the summer, flowers as well. The entire town is incredibly clean—indeed, I have never seen the like. People are constantly employed with sweeping streets and removing the filth. The residents also appear healthier and stronger than any other place I have been. I wish I had had more time there, so I could give a complete description; but one cannot see much in half a day.

In the afternoon we sailed towards Southampton, where we arrived about eight the next morning. Here we took several Englishmen onboard, including a planter from Ceylon with a wife and children. They had a Hindu nanny, and it was quite amusing to see how devoted the black man was to the little, white children. I believe the children cared more for him than for the sallow skinned parents. We were here only for a few hours and sailed away once again. It is particularly beautiful between England and the Isle of Wight. Large, beautiful fortifications are built on both sides to protect the entrance.

On Thursday morning the 12th of April, we arrived at the Straits of Gibraltar and were, naturally, up early for the view. But, as it was still dark, we could only see the outline of the Spanish and African coasts, which looked quite barren and empty. Even though it is a dangerous place, the lighthouses are thinly spread. When we were a good way beyond Gibraltar, day began to break and the Spanish coast came into view with its snow-covered mountains. I have never seen such a beautiful sight. Imagine the Mediterranean as smooth as a mirror. The small, white fishing boats with two, sliding gunter's sails appear at a distance, like large bats. They lie nearly still, when the wind is slight. The sun slowly rose over the top of the mountains, guilding their cover of snow with its rays, which is reflected again on the water, so that it appears miles deep with a golden bottom. The mountains rise up from the sea to the clouds. Towards the bottom they are covered with small bushes of a brownish colour. A light blue fog lies from the foot of the mountains almost up to the snow-covered top. When the clouds are over the mountains, it is difficult to say where the clouds begin and where the mountains end. The mountains are 5 mil [twenty-five English miles] from here, but they almost look like one could jump in the water and swim to them, the air is so clear.

---

[1] Østergade in Hjørring is 17–20 metres wide (it varies a little along the street).

[Letter printed in VT] 3rd August 1888
(Continued)

For three days we had the Spanish coast in sight. We then steamed by that part of the French coast with the Alps lying between France and Italy, whose ice fields sent such a cold breath out over the sea that I had to get out my winter coat, which I thought I had said farewell to for a long time. On Sunday the 15th of April we arrived at the city of one hundred churches, the enchanting Genoa. On the mountains surrounding it on three sides, there are lots of grey, fortification walls, with their black, cannon barrels, like searching eyes, turned towards the sea. Otherwise the magnificent city lies nearly hidden from the seaside, so that when estimating its population or size, one can easily go wrong. The city, in fact, has 180,000 residents, but, from the seaside, it does not look like more than 10,000. It lies picturesquely up on mountain slopes. On the top of one, you see a wealthy man's beautiful villa. The trees were in full bloom. The heights were lush with oranges, rows of grape vines and cedar, and the white houses shone invitingly through their leaves.

We were eager to see this enchanting city as it peeked out from the mountains. We went ashore in a gondola

[a tourist trip around Genoa with visits in churches etc. has been omitted].

There was enough to see here to fill months, but I had only a day and a half. We had to quickly return to town. On the way we passed lush grape mountains, while, closer to town, there is a fairground with the type of performers found all over the world, along with happy families with picnic baskets along the grass-green slopes. There is a completely charming view over the sea, the town and the harbour with the beautiful ships from all over the world. In many parts of the town the houses are built up against the mountains, so that from the second story, you can step right out onto the slopes, which have been levelled and converted to gardens. Tired out by our tour—after visiting a restaurant—we went into such a garden to enjoy a refreshment. The garden was full of oranges and flowering hyacinth bushes. Up against the wall, ivory wrapped around wonderful, marble figures, and there, in the middle of the garden, a fountain splashed and goldfinch played.

What a change when I consider that three weeks ago, I was ice skating on Svanelunden in Hjørring, in which town fat Madam Bollerup and the beautiful Nikolaj and the other snow figures collected money for the freezing poor. Now I could throw away my coat and enjoy life in this green, flowering climate, where oranges hung so close over my head that I only needed to stand on a chair to pluck them. It was with a heavy heart that I returned onboard.

[Letter printed in VT] 11th August 1888
(Continued)

It was with a sense of loss that I left Genoa, as much here was new for me. One sees little art in America, and you know that I have not had much opportunity to see it at home. We lifted anchor in the afternoon and slipped out of the harbour, while music played. The town became smaller and smaller, and I remained aft until the last house slipped from sight. However, I promised myself that when, in time, I return home, the journey would go that way. So I will come home by railway through Italy. On the 18th of April we passed the formerly fire-spitting Stromboli. The mountain rises abruptly out of the sea like a large stone. The top, however, was shrouded in a huge cloud, which hid the height of the old fire breather. At the foot lay a small village with an apparently large grape production. Its green gardens covered the slopes. At noon we steamed through the Straits of Messina. The high mountains and the large orange groves created an inviting picture on the Italian side, but Messina has a barren appearance close to the sea.

After three days of sailing we came to the Egyptian town of Port Said, on the edge of the vast Sahara Desert and close to the Holy Land. It is barren, desolate and as hot as a baking oven. One almost never sees a tree. You complain up there, where the last, green beech will soon bloom; you should see this place. Nearly the entire population is Arab. Men and women have very similar clothing. The married women veil their faces. A short, brass tube with three rings is placed on the middle of the forehead, to which the veil is attached.

All the houses are made of large bricks and plaster, since there is neither wood nor stone here. There is also no drinking water. It comes from the south through heavy iron pipes and runs out into large, walled tanks. From there it is transported around the town in sacks of cowhide, hair side out. Filled with water, these sacks are placed on the back of a donkey, which, with blinking eyes under the baking sun, carries them to their destination. Donkeys and camels are irreplaceable here, as they are the region's only means of transportation.

Have you seen a worm fall down into a school of fish? That is about what it is like when one comes into a crowd of donkey drivers, who flock together, when one comes ashore in Port Said. When we first came out to ride the next day, we rejected them—but they are not easy to get rid of. In a language sounding like the tower of Babel had just collapsed, they yelled at us. "Here Mister, gut æsel, Berliner æsel! Bismarck æsel!" ["Hey Mister, good donkey, Berlin donkey, Bismarck donkey!"] screamed another competitively. But, when not even a long-eared Bismarck donkey could tempt us, they finally left.

In the European quarter we visited a concert hall. All the musicians were

Bohemians, but of different nationalities. Nevertheless, you can imagine my surprise when a lovely little girl suddenly began to sing: 'There is a Lovely Land' [the Danish national anthem], while the music accompaniment softly played the well-known, beloved melody. It is perhaps superfluous to note that I beat on the foot of my glass in applause. Thereafter we went to see Arabian dancing—one of the strangest things I have seen. The music for it is difficult to enjoy. One of the instruments resembles a large water carafe with a bladder attached where the bottom should be. One musician drums on this monotonously and incessantly, and an equally deadly boring bagpipe gasps along with it. The dance is in good harmony with the music. The dancing girl turns herself on the same spot, while she swings two scarves over her head. The stomach and hip movements are particularly noted. Periodically she stops abruptly on one spot and remains standing like a pillar for five minutes, moving only the stomach. It mostly reminded me of the position a sick dog takes when it is trying to throw up and pulls its stomach in between its ribs. When she first stopped, I actually thought she was sick. The cheering, however, taught me that it was thought to be beautiful—charming. Young men who are obsessed with the enchanting dancing girls told about in *A Thousand and One Nights*— remember the sick dog!

Port Said is the most immoral city I have ever seen. It is as if all the smut of indecency found all over the world among savages and the civilised comes together here. On public streets pictures are sold, the contents of which I will not even hint at. There are dancing theatres where women do not even have a shadow of clothing. Side by side, there is, of course, a lack of cleanliness that is more than swinish. In any case the old man who taught me agriculture would have growled loudly if the sty had smelled the same as the streets here. There is also a great deal of poverty. Everyone begs, and it is not unusual to keep the most pressing away with a chair, which is nearly indispensable.

The Arabs here do not need to be so dirty, since the water is pleasant; and, as a rule, they swim quite often. A very small little boy here dove to the bottom after some English coins I threw in the water. And, for 25 øre, he went to the bottom on one side and came up on the other side of our ship, which draws 25 [Danish] feet under water. That was quite a feat!

[Letter printed in VT] 14th August 1888

On the 23rd of May we sailed into the Suez Canal, which, to me, does not look wider than Ryaa.[2] When I sat in the cabin, it looked like we were riding

[2] Ryaa is a river in Northern Jutland, Denmark. It is approximately 30 meters at its widest today. Until the end of the 1920s steamers sailed up the Ryaa. The Suez Canal was originally rather narrow (22 meters at the bottom of the canal), so CL's comparison of the width of Ryaa and the Suez Canal in 1888 is probably rather accurate.

in a train, because, through the windows, we saw only the land quickly gliding by. Three English steamers that we needed to pass made room by towing one of the ships to the side with hawser and winch to make space for the other. Signal stations along the bank constantly announced oncoming ships on the narrow waterway between the world's oceans. The canal is also constantly being worked on. For example, a wall to prevent the water from washing the sand down was being built. The project employs about 200 men with 100 camels. The latter hold their heads high, as if they were directors of the National Bank, willingly carrying stones to the work site. Here they patiently lie down to be unloaded and are then returned for a new load. The small, dark, Arab children came to beg as soon as they spotted our ship. We threw French bread at them, which a clump of children immediately fought over. I can understand that they are hungry; one becomes thirsty just by looking at the desert coast along the Suez Canal. Here there is literally nothing but hot sand, which, when the wind blows, fills the throat and all other openings. And, when the sun shines, it burns so that it is not complete fantasy when the boatman sings: "Oh all the eels we catch were fried so they were seaweed." In any case the heat is unbearable.

One comes through the two bitter lakes before reaching the Red Sea. I do not know if the water here is bitter, but I did not get through them without a bitter memory. In the large lake we met the steamship *Slesvig*, and, in honour of that, the Germans in our group wanted to sing 'Slesvigs Pris' [Schleswig's Praise].[3]

But I begged to be excused from the honour of singing it on the German side. I have met too many people from Schleswig who were warm Danish patriots to do that. The captain noticed me, gave me a small smile and said: "Well, I thought that you would not go along with that idea. You should know that you are not the only one. I would not sing along either. Sit down and we can drink a cup together for old Denmark, while they sing up there." He is from Schleswig and speaks Danish as well as I do. We found each other the first day and have been friends since.

The next day we were in the Red Sea, where I, naturally, saw with great

[3] The song mentioned was probably 'Schleswig-Holstein Meeromschlungen' which was the 'national' anthem of the two dukedoms of Schleswig (Slesvig in Danish) and Holstein (Holsten in Danish). Especially Slesvig was a nationally contested area. Until 1864 both dukedoms were part of the Realm of the King of Denmark (but never a part of the Kingdom of Denmark). After the Second Schleswigian War in 1864 The Kingdom of Denmark was forced to cede Schleswig and Holstein to Prussia and Austria-Hungary. This led to a loss of more than one-third of the population of the Realm, and one of its most economically advanced areas. The loss was a shock. People will still refer to '1864' today, speaking of the reorientation in Danish politics, culture etc. in the wake of the lost war in 1864. The Northern half-part of Slesvig is today part of the Kingdom of Denmark (ceded back to Denmark in 1920). For a short introduction in English to the history of Denmark including the 1864-syndrome see Knud J.V.Jespersen, *A History of Denmark*, Palgrave/Macmillan 2004.

interest the place where the Jews crossed over. Further into the day we also passed Mount Sinai of the Ten Commandments. That God had to let it rain manna and allow water to burst forth from the earth, I can easily understand. It had to happen in a supernatural way, when He brought his Jew's there, because only sand and stone cover the earth. Birds do not fly through the air, and one hardly sees an insect. Everything is barren and empty and cloudless under a torturing, hot sun.

On the 20th we came to Aden, which is strongly fortified by the English. It is stony and unfruitful, cursed even more, since the well water is so bitter that only rainwater is drinkable. When it occasionally rains, which comes in downpours, the water is collected in large, natural stone basins constructed by the English, who have built a small, cosy plantation around them. This provides such a relative beauty in the middle of the desert that it is a strong argument for the tree-planting movement[4] that, once one has felt the terrible desert here, will not soon be forgotten. There are about eight basins that, together, can receive six million gallons of water, sufficient for people for several years. The water is carried to the town by small donkey and camel caravans loaded with cowhide sacks as in Port Said.

Every morning there is a market here in the town. In the background are mountains. At their feet are long, white buildings with arched walks. In front of them, on large, stone fenced squares, people stream together with hundreds and hundreds of camels, which, in small clusters, come wandering from all directions. Every type of trade has its own square. On one side are the water sellers, whose water-sack loaded camels resemble a bundle of balloons. Then come the same beasts of burden carrying brushwood and other firewood, every piece of which must be carried here, since, as noted, the area is treeless. Then the 'farmers' bring their meagre produce. Then come the fishermen's camels with their fresh-caught fish, and, finally, we come to the square where the camels bring meat to market. While one otherwise only sees a white-clad figure here and there between the loaded animals, the meat market is a veritable whirl of people of different nationalities, Indians, Negroes, Arabs, Persians and Germans. But the language is mainly Arab, and the dress mostly the same. It consists of a long piece of cloth that, wrapped, makes up pants, shirt and head cover. There are often pretty faces hidden under these folds, who would be pleasant to meet alone and unarmed. The culture is perhaps described best by explaining that a young girl's preparation for marriage here is a good number of severe blows.

---

[4] The tree planting movement (skovrejsningsbevægelsen) has its roots in the enlightment and its belief in mankinds rational command of nature. Private, but supported by state legislation, new forests (plantations mainly of pine trees) was planted on the sandy soils mainly in Jutland as part of a land reclamation project. The tree planting movement gained momentum in the 1870's and 1880's (after the loss of Schleswig and Holstein in the Second Schleswigian War in 1864).

If she receives them with a mild face, it is proof that she can accept for better or for worse.

Labour is not very expensive. At the hotel everyone has his own servant, who fans him and brings him cooling drinks when it is warm; and who can be used as a pack animal when one goes shopping in the town. The buildings here, because of the rich insect life, contain the least possible wood. Swarms of cockroaches find wood such an excellent material for their holes that they cannot be exterminated. The floors are of stone and the beds of iron with mattress and pillow but no cover. A quilt would be a joke here, where all gasp from such a burning heat that none of us, who are not used to such heat, could sleep the first night.

Arabia Aden, 1st May 1888

Dear Brother-in-law and Sister,
Just a few words to let you know that I am alive and well. Here I am now in what I believe is the hottest place on earth. The town lies down between mountains of stone, and, when the sun shines on them, they become so hot that the air coming from the mountains is almost burning. I have reduced my clothing to a pair of pants and a jacket. That is all I have on now. Shirts and underwear are impossible to tolerate. It is also much warmer than Zanzibar. The Germans I am travelling with are in despair over the heat and must have ice in the water when they bathe in order to cool off. I wonder what they will say when we come to the African jungle; then they will not have ice in the water.[5] Here it sometimes does not rain for three to four years, and, in order to have enough water, large water tanks have been constructed between the mountains. I believe there are six, and the largest can hold approximately 3 million gallons of water, which is transported to town by camel or donkey. The Negroes and the Arabs here wear almost no clothes. We are staying at the best hotel in the town, and our waiters have only a pair of swimming pants[6] on, while they serve the tables. I am sending you a number of pictures of the different people who live here and of the town and region. I have written on the back of each to help you. Because of this horrible heat, I do not feel like writing anymore, and I would,

[5] Modern refrigerators and freezers were invented and developed in the 1860's and came into use industrially from the 1870's onward. Ice could be manufactured from sea water as well as sweet water. Sultan Barghash had an ice factory built in Zanzibar in the mid-19th century (Eric Gilbert, *Dhows and the Colonial Economy of Zanzibar*, Columbus: Ohio, Ohio University Press), 2004.

[6] Swimming pants is the direct translation from Danish. A language history dictionary (*Ordbog over det danske Sprog*, Gyldendal 1946) simply explains that swimming pants are the same as short pants. So in this context it must simply mean that the waiters wore short trousers.

therefore, wish that you would send this letter to mother, so that she can also see that I am well.

Best wishes to all.
Your affectionate brother and son-in-law.
Chr. Lautherborn

Deutsch-Ostafrikanische Gesellschaft (German East Africa Co.)
Zanzibar, Africa via Brandisi [no date]
In Aden I had to say farewell to all of my travel companions with the exception of a few who were continuing to Zanzibar. One always becomes so well acquainted on a ship and builds better friendships there than any other place. It was therefore with a heavy heart that I said goodbye to my friends. I would rather have gone ashore without saying anything, but, on the other side, it would have looked strange; so I thought 'Take Courage Antonius' and went through the unpleasant ceremony of saying goodbye to each one. I listened to all of their good wishes, got their addresses so that I could write and tell them about Africa, shook, in conclusion, my captain's hand and went down the gangplank with a pressing sensation in my throat, like Nicolai in *Nøddebo Præstegaard*[7] when he was choking on the potato.

From here I had to sail with the British-Indian steamship *Java* to Zanzibar. It was quite a change to come from the well-maintained German ship to the British, where all in the cabin had a greasy appearance. The cabin swarmed with cockroaches, so I did not get any sleep the first two nights because of them. One can, however, get used to anything, and so it went here too. After I stayed awake for a time in order to keep watch on them, I became so sleepy that, even when they ran over my face and down my back, I continued to sleep. One of my companions did not sleep the entire voyage. He was so afraid of the cockroaches that, when we arrived at Zanzibar, he looked like death warmed over. The sailors ate anywhere on the ship. Their food consists exclusively of rice and curry. A large, round, tin plate, about three feet wide, is filled with rice. In the middle there is a bowl with a mixture of butter and curry. Six to seven men sit around the plate on the deck, make small, round balls of the rice with their fingers, dip it in the bowl of curry and swallow it. Knives and forks are a luxury for these people.

[7] *Nøddebo Præstegaard* is a still popular Danish stage play enacted for the first time in 1888. It is about three brothers who during Christmas holiday visit a vicarage (*præstegaard* means vicarage—Nøddebo is the name of an existing village). The vicar has two young, beautiful daughters whom the brothers of course fall in love with. The character 'Nicolai' is the youngest of the three brothers in the stage play and like in fairy tales the youngest brother is the lucky one who in the end marries the princess (i.e. wins the heart of the vicar's daughter).

Most of the passengers were Indian, the most swine-like people I have ever seen. Their clothing consisted of a piece of cloth from four to five alen [yards] long and a little more than an alen [yard] wide, which was wrapped around the waist, so it reached a little over the knees. Their hairstyles are quite peculiar. Some have all of their hair shaved off. Others shave half from ear to ear, so that only the hair in back remains. Others have their hair clipped in the shape of a cap, and, in the middle of the head, they allow the hair to grow so far out, that it looks exactly as if they had a black cap with a long tassel in the middle. Still others leave only a little, square piece on the forehead, which is cut very short. All were of the Moslem faith, and, morning and evening, one could see them at their prayers, when they fell on their knees and struck their forehead against the deck. Before they begin their prayers, they always wash their feet, while the face is usually neglected. The result is dirty faces with clean feet. They carry all their drinking water with them, because they will not drink the water on the ship, which is, in their opinion, unclean. The priest must first say a short prayer over all water that is drunk. They, therefore, carry a lot of water with them, when they go on long trips, so that they will have enough until they reach the next priest. A quite small bowl, somewhat in the form of an hourglass, is used to drink from, but they do not touch their lips with it. The head is thrown back, the mouth opened, and the water is poured in.

From Aden a number of sheep, chickens and ducks were brought for the Europeans. One could easily see that they came from a place where almost no grass grows, since both the poultry and the sheep were so thin that, when served, one could find little more than skin and bone. If there was any meat, it was so long and leathery, that I believe one could stretch it from one end of the cabin to the other. The ducks and the chickens got a little old rice to eat, while the poor sheep starved every day. They were so hungry that they ate straw mats and old cloth to dampen their hunger. They were given water every day. I asked the sailors why they did not feed them. They answered that they were going to be slaughtered, so there was no point in feeding them, since the stomachs were thrown away. I have a feeling that, by the end of the voyage, they did not need to slaughter them, since they were probably dead of hunger.

After eight days of sailing we came to a little town by the name of Lamu on the East coast of Africa. Since Lamu lies a ways inland, we dropped anchor outside a little village whose name I have forgotten. We sailed inland with an Arab, who invited us to his house where we sat on straw mats. While we curled up our legs under us, a little Negro boy carried a pipe around to each in turn. Everyone had to take a drag on the peace pipe. Tea was then served in a European tea set, which a German had given him. We drank the tea quickly, because our legs had begun to sleep and throb in positions they were not used to. I know that it was almost impossible for me to hold out longer. I mostly wanted to stretch my legs out into the teacups sitting on the carpet between us. Since the others were just as tired as I was, we first had to stand up and hold on to the wall, until the blood began to circulate again in our legs enough for us to use them. When we had regained the use of our legs, we said farewell to the Arab and started for Lamu.

The way there led through a very large palm tree forest, where coconuts hung by the thousands over our heads. Since it was very warm, we paused under the shadow of a large tree to cool off. A Negro with a pair of swimming pants on brought us palm wine in a kind of hollowed out gourd. To take the wine from the tree, a hole is bored in the top of the trunk into which a pipe is stuck. A bucket is hung, and the sap runs out through the pipe. This is taken down each morning, while it is cool. The wine [*tembo*] tasted very sweet, almost sickeningly so, but it is supposed to be, when aged, very intoxicating. After an hour we arrived in Lamu. Here, for the first time, I saw the disease called elephantiasis, which is a very good name since both the leg and the foot swells so that it looks exactly like an elephant's. I have seen the leg at the ankle swell up to three times normal. I hope that I will be spared this horrible disease, dragging around twenty-five pounds of hardened meat on my legs.

We were a great surprise in the town, because Europeans are a rarity here. In less than ten minutes after our arrival we had a retinue of approximately one hundred persons, all with a weapon of one kind or another. In the middle of the town was a large, square space used as the town square. It had banana plants growing in it. We stopped to admire the beautiful plants, and were soon encircled by such a mass of people that we thought that we would never come out again. Some of them had very long spears. I went up to study one closer, which was truly a masterpiece of craftsmanship. The spearhead was very long and sharpened with a groove down the middle. The shaft was round and smooth as if it had been made with a lathe. About a foot from where the shaft was connected to the spearhead it was wrapped with brass wire. When one held it, the entire shaft vibrated; and the spearhead was as sharp as a razor blade. I involuntarily thought how unpleasant it would be to have such a spear go right through one's liver, but, as they were all peaceful people, we really had nothing to fear.

The women all have rings in their noses. They cut a hole in the lower part of the ear, where a sharp tipped block of wood is set. One continues to press the block in, until the ear becomes as big as it is possible to get it. The block is then cut, so that only a disk as thick as the ear remains. For beauty's sake the women paint their eye lashes and under their eyes dark blue, yellow, white or green. Dress consists of a piece of cloth, which reaches from the waist to knees. Small, newborn children are carried in a piece of cloth on the back. The mother works every day with her child on her back until they are big enough to walk. When the Blacks have a child, their breasts become so long that they nearly reach the navel. When a little child becomes thirsty, the breast is cast over the shoulder and the child nurses, while it hangs on the back.

A German lives in Lamu, so naturally we had to visit him. He had been in Africa for about fifteen years and entertained us with stories about the animals and his hunting expeditions for a couple of hours. When we stood up to leave and see more of the town, a Negro, who spoke English, was our guide. He told us that there was a big place in the town, where one could sit down and drink a bottle of wine. We went there, but the honoured innkeeper apparently has little business, as he was not there. When he came, his entire stock consisted of three bottles of red wine of which we drank two mixed with water. We sat in front of the house in the shade of a large tree. The entire street was packed with people there to look at us. When we clinked our glasses before we drank, they howled to the sky with laughter and slapped their thighs, and poked each other with their elbows in order to express their amazement over such a manner of drinking. I believe that both the Blacks and ourselves were greatly amused with each other. Many witty remarks were made on both sides, which, however, were not understood. It was only through our guide that I learned how much the Negroes had been amused by us.

Since, however, it was nearly time for our steamship to sail, we went down to the bay and made an agreement with a boatman to take us out to the ship. But, when we got into the boat, the current was so strong that, after we had been in the water for half an hour, we found us half a mil [3.2 miles] from where we should be. We soon realised that we would never reach the ship that way. We, therefore, had the boatman set us ashore, and went by foot to the little village we had been to in the morning. As it was very hot, and we had to walk quickly to get there in time, we were all very hot when we reached the village, where there was only one canoe available. The ship was about to turn in order to sail out, so we had little time to waste. We crammed into the canoe and shoved off from land. We had not, however, gone more than a couple of hundred alen [yards] before a large wave struck the side of the canoe and turned it bottom up. We fell, head first, into the water. Those who could swim were kept busy rescuing those who could not. Two Catholic priests were in a dangerous condition, because, with their long cassocks down to their ankles, they could not move their legs when they were wet. We got them to hold on to the overturned canoe. A fat Jew, who could not swim, became so frightened when the canoe overturned, that he screamed like a wild man for help. His fear was, however, ungrounded, because he was so fat that he floated on top just like a cork. But we got him to the boat, where he held on, while we who could swim pushed the boat toward land with eight people hanging on to the sides. The ship had seen our difficulties and sent a boat out to rescue us, but since the current was so strong, we reached land before the boat came. We were, however, glad to be able to come into a good boat, which we knew could stand up to the waves. We reached the ship without further trouble. We had all purchased a lot of different fruit, all of which went to the bottom when the boat overturned. I, and one of my companions, had bought a large bunch of bananas along with a couple of the natives' handmade knives, all of which went to the bottom.

The Jew, in his fright, lost a quite good telescope and a gold-knobbed walking stick. A monkey he had bought had somehow got away from him. What became of it I have no idea of, as I had enough to do saving myself and the others. As a consequence of the incident, however, we all got a fever, which many still had when we came to Zanzibar. Since I was well acquainted with this illness from my Texas days, I cured myself and was well again the next day.

The next evening, when the lanterns were lit, flying fish came up on the deck in such numbers that it was almost dangerous to be there. One flew right into my nose, which smarted for a long time afterwards. If it had hit my eye, it would quite probably have blinded me. But the flying fish performance soon ended. After just one and a half minutes it was over. It was the only time during the voyage that I saw them fly up on the deck. They must be able to fly pretty high though, because the sun awning was between 25 and 30 fod [feet] above the water; and we got a couple of dozen fish from there, after they had stopped flying. The next day we ate fried flying fish, which tastes a lot like herring.

On the 16th of May, at noon, I arrived at Zanzibar. The voyage along the coast with its rich, green colour, and the large palm and mango trees, which grow all over the coast here, was beautiful. We arrived at noon, but the large clock showed only six. I did not understand, but learned that Arabian time is different from all the others. When the sun stands up, it is twelve; when it goes down, it is twelve; and at noon it is six. The company's boat was there, when we arrived. We were taken ashore and presented to the various Europeans who represented the German company here, given a good dinner, and, at three o'clock, when we were finished, went out in the town to see a little of its wonders. Zanzibar is, after all, considered by the people of the East Coast to be the world's largest city. It is built exactly like all other African towns I have been in, with the buildings so close together that the streets are only four fod [feet] wide, and they run in all possible directions. Consequently it is completely impossible for a man to find his way out of the town, after he has come in. The buildings are only one storey high with clay walls and palm leaf roofs. It is, however, only the Negroes' houses that are built that way.

The wealthy Arabs and Europeans all have brick houses. The Sultan's palace is the most beautiful building in town, but, for a European, very ordinary. It is whitewashed and four stories high. It has a zinc roof. It is square, with a balcony that goes around the entire building on each storey. Three, three storey buildings are close to the palace, connected to it by a covered passageway. This is for the Sultan's harem. In front of the harem is an artesian well. Around the well there is a four-foot high wall, which holds the water until it reaches a certain height. It then runs out through countless small pipes, from which the Negroes, all over town, get their water. A ship of stone painted grey is in the middle of the basin, making it look like it floats on the water. An English machinist, who had brought a machine from England for the Sultan introduced me to one of the ministers. I asked him if it would be possible for me to see the harem. He appeared to be quite shocked by my request and rejected me flatly. There is a very stiff and reserved tone between Europeans in Zanzibar, which I do not care for at all. But, since I will not be staying here, it makes no difference.

In front of the palace there is a large area paved with cement, where the Negroes, on Friday, dance their wild war dance in honour of the Sultan. There are many slaves here. They, however, do well and are well treated. The Arabs have their wealth in slaves, and a number of them have at least 500. A slave can be bought cheaply. A man can cost from 700 to 800 kr., and a young girl from 400 to 500 kr.[8] That is very cheap, but, when the large caravans come from the African interior, slaves are even cheaper—a man can be bought for 200 and a young girl for 100 kr. An English warship is here to prevent the slave trade. It thrives pretty well anyway, even though I must say that the English capture many slave ships and set the Negroes free.

The place I am to set up a cotton plantation is a little ways inland from Pangani close by the river, which I reached, without further adventure, on an African sailing ship, from which I write this letter. Because, however, I have written so much that I am now tired, I will tell you about the life of the Blacks here another time. I know it will interest you. Even though I have been used to Negroes from Texas for eight years, I still find their life and behaviour extremely interesting. Give my greetings to all the good people I have known in Hjørring. I have much to do every day. I am studying Swahili intensely and have begun to speak it a little. But it is a strange language with absolutely nothing in common with other languages. I hope that you got the pictures I sent you from Aden? I have begun to collect different things such as coral, turtle shells, and snake skins. I found an oyster shell that was nearly two feet across and weighed, I believe, 40 [Danish] pounds. It stands on three legs in my room, where I use it as a wash-bowl. It is very inviting to set one's hands down in the clear water, because the shell inside is looks like mother-of-pearl; and the water seems to become clearer, when it is put in.

---

[8] 500 d. kr. equals approximately 16.000 Danish 1998 kr. 16.000 d. kr. were about the same as 1.500 British Pounds in 1998. The average yearly salary for a farm hand (man) in Denmark in 1888 was 404 d.kr. (see Jørgen Pedersen, *Arbejdslønnen i Danmark under skiftende Konjunkturer i Perioden ca. 1850–1913.* København 1930, p. 112–113).

A friendly greeting to you, Caroline, and
to the children from your devoted brother-in-law
Christian Lautherborn

[Letter printed in VT] 14th December 1888

From the Coast of the Dark Continent
Vendsyssel Tidende's correspondent—before the uprising:
When I say that there is not a single nail in the house where I live, it sounds nearly unbelievable to European ears. None the less it is true. Our entire 'residence' is bound together with cords twisted from palm leaves. Here it is better to be a rope maker than a smith, at least as long as it lasts. The frame consists of bamboo poles bound together, and the walls are thick, woven palm mats bound to the poles. It feels a little strange to lie down against such a thin mat and hear the wild animals roar in the forest nearby. The window frames are attached in the same, simple manner as the rest. The windows themselves are replaced by thick blinds, which are rolled down when it is windy. The roof consists of layers of bundled palm leaves laid like our Danish thatch roofs.

This light building stands on a five fod [foot] high, walled elevation and is extremely pleasant and cool. It is 50 square fod [feet], and, with rafia mats, divided into two bedrooms, that take half of the space, and one, large room, where we eat, read and have our office. Of our three desks, one stands by the window opening, so that, through this, we can hand money out to the Negroes without coming into close contact with them. The reason for this procedure is not that we lift our noses at them, but our noses do play a part. The Negroes rub a kind of grease on their bodies to make the skin soft. When they work they sweat a great deal. The sweat and the grease combined create such a horrible cover on them that they stink like the plague. Indeed it could be used to poison rats.

Our room is quite comfortable and well supplied with chairs, in which one can almost recline. Rifles, shotguns and ammunition bags are attractive wall decorations as well as being practical to have nearby. Purely as decoration we have native spears and shields along with a couple of Negro skulls, which we keep because of their odd, I would say, inhuman shape. I could not sleep the first couple of nights because of the monkeys' horrible screaming and crawling on the house, the roof of which they have chosen as a dance hall. I got used to this quickly. But there is one thing I cannot get used to, the horrible snakes. During the day they hide between the roof's palm leaves; at night they slither out to hunt, and I often have them as guests in my room. The third night there I awoke to a great disturbance between the dry palm leaves and a fall of heavy body down against the table next to my bed. At first I thought it was one of the apes that had slipped through and rushed to light a candle in order to get it. But, to my horror and surprise, I saw that it was a quite large snake, which, with its small gimlet eyes, was staring at me and the candle. With the speed of lightning I grabbed my pistol and aimed a shot at it. I was lucky enough to nearly cut the head from the body and, in the next second, had to laugh at my brave colleagues, who had rushed in their night clothes and now stood and gave Pierrot's fright[9] very disappointingly. After I had dragged the snake outside, we settled down again, but since then I have slept with a light. It keeps all such creatures away. As a rule they do not come out when a room is illuminated, and, in any case, one comes into a defensive position more quickly.

I am happy about my beautiful cotton field and look forward to the first harvest with excitement. It can, however, be threatened by many dangers, which I do not even know yet. The Swahili language also causes me great difficulty. It does not resemble anything I have heard, but I am studying hard and I shall and will learn it. It is strange how the small words help, when there is no other way.

[9] Pierrot is a character in a stage play/dumb show. Because it is a dumb show the meaning has to be conveyed by expression of the face or science of gestures alone.

CHAPTER FOUR

THE WAR IN EAST AFRICA: PANGANI AND BAGAMOYO

[Letter printed in VT] 20th December 1888

From *Vendsyssel Tidende's* Correspondent
(Continued)

In the beginning it looked like all would go reasonably smoothly [in Pangani].[1] Since the Sultan's 'soldiers' had only very little gun powder and shot, we did not have much to fear; and we would be sure to keep alert, for we had no faith in these villains.

Unfortunately they soon received help. After a couple of days [on 3rd September 1888] a guest arrived, as unexpected as it was unwanted, a *dhow* with an entire cargo of ammunition. With a real horse thief's smile the crew claimed it was from an Indian. I do not doubt that the sender was the Sultan by God's grace whom, as little as a Christian diplomat, is afraid to manage with a lie in a narrow phrase.

When our chief, Emil von Zelewski, learned what a fateful cargo the *dhow* carried, he gave a strong and determined order that the ship should sail back to Zanzibar immediately. During the existing unruly conditions he did not want a large supply of gunpowder here. But, when the captain gave orders to raise the ship's only sail, the soldiers[2] suddenly came to life. Threateningly they declared that they could use all the ammunition and would not allow it to be sent away. It was clear that there was a conspiracy between the soldiers against us.

As it was Sunday, I was in Pangani and saw how their courage and enthusiasm rose when they got the ammunition in their hands. Even though they were forbidden to shoot in the town, they began to press in near our house, where they shot off their guns. True enough, they shot in the air, but with an expression on their excited faces as if they had the greatest desire to hear the bullets penetrate our flesh. Our chief [Zelewski] asked them coldly and determinedly if they were aware that he had forbidden all shooting. With

---

[1] Lautherborn severely underestimated the negative impact of Zelewski's behaviour and paints a rather more optimistic picture than the facts suggest. See chapter two (The African context).

[2] Called *viriboto* in Swahili, they were the irregulars who served in the Sultan of Zanzibar's garrisons. They were low-status Arab mercenaries who generally practiced Sunni Islam. (Glassman 1995: xvii).

wide grins they laughed in his face, and had the nerve to aim at us with their loaded guns. In such a moment self-control was required to not open fire on them. We stood there in Pangani, eight Europeans, and were ready to dare a fight with Black, provoking scoundrels. But the chief had orders not to shoot before he was attacked. It was a difficult order to follow, because it is easier to strike than to calmly tolerate insults of that type. However, it was done. The company's steamship, which, by chance, was in service, was quickly sent at full steam after a warship. As one can perhaps understand, the chief had nothing against quickly getting rid of the local soldiers. He wanted them 'fetched' as soon as possible, but armed force was absolutely necessary for this.

In the evening I walked back to the plantation in the belief that all would soon be in order and the people calm. But, after three days, I received a letter from a friend in Pangani that read approximately as follows:

> Satan is loose here! This morning one man aimed a pistol at me. Our lives are no longer safe. A huge number of savages from the interior stream into Pangani to kill the whites. The Arabs have spread the rumour among them that we will begin a slave trade on a large scale, do away with the Moslem faith and force them to become Christians. The Blacks are furious, and right now we do not dare leave our homes. It is certain that they will soon attack us. We advise you and Hr. M. [Lautherborn's assistant] to come in for a couple of days, because, when it breaks loose, it will be best that we are together; because we will be better able to defend us. You cannot make a stand in your thin house of palm leaves. Come this evening if possible, as we await you by the hour. If you stay out there, it could cost you your life.

As soon as the sun had gone down, we grabbed our guns and tried to sneak into town in the dark. It worked, even though we found a quite large group of armed Blacks by the river, who tried to prevent us crossing the water. But we raised our guns and aimed them at two oarsmen. This got them to work at getting their light canoe, with a foaming white stripe following, quickly to the other side near to the German house, where our friends received us with great pleasure.

The next day [5th September], the warship was to arrive, and, with a relieved sigh, we saw out to sea a little, white dot, which gradually became larger and larger. With binoculars to our eyes, we did not relax a minute before we discovered it was not a warship, but the Sultan's steamship, *Barawa*. In amazement we asked each other what it wanted, and we soon found out. When they got close enough to shore that we could make out individual persons through our binoculars, we saw our governor-general [Ernst Vohsen, chief officer at Zanizibar for DOAG] with the two interpreters and the captain go into a small boat to come to us. A roaring salvo of gunfire broke out from the beach as soon as they were close enough. In the next moment the oarsmen sought cover from the bullets, which partly flew over them and partly struck the boat. When they stopped rowing, they drove toward shore, where the Blacks, in bloodthirsty excitement, prepared to move in and kill the Whites. The captain quickly drew out his revolver and threatened to send a bullet through their heads if they did not immediately take the oars. This helped! They sat on their knees in the bottom of the boat and kept their heads below the railing, so that you could see only the arms working with the oars stuck outside. In this equally clever and laughable manner they rowed until they were outside of range. They then set sail and, with great speed, sailed toward the ship, which immediately sailed away and left us with the roaring savages. With a canoe that was passing by they sent us a message saying that they sailed away to find the German warship; all had been taken from Zanzibar to receive Admiral Deinhard.[3]

As soon as the Blacks saw the ship sail away, they became almost uncontrollable with fighting spirit, because they thought they had repulsed an attack. Now they would immediately attack and slaughter us. But a couple of older and calmer heads recommended that they wait a couple of days, because there was another job that had to be done first. This was to be done the next day. In the evening the Blacks held council, which ended with a wild dance to the thundering tones of the large war drum, which could be heard for miles.

At ten the next morning [6th September] about 3000 Black men assembled a mile from our house. They began to march towards us, while beating the war drums and blowing on long antelope horns—the most Judgement Day, horrible sound I have ever heard. When we saw them rolling up, we were certain that they had come to kill us. I can dare say that the minutes that passed until we knew their real errand would have driven many men mad. With the most horrifying noise they came. It looked like all the drums, horns and other barbaric instruments to be found in Pangani were in use. They had painted their faces in all possible colours, some with large, green eyes, others green with white eyes, again others with yellow eyes, yet others yellow with red eyes. Their battle clothes consisted of black monkey skins ending close to the throat and hanging down to the knees in front and back and equipped with bells on the bottom, just like the ones we use home in Denmark, when we go 'to town' in a sleigh. Below the knees they had also tied a row of large bells. On their heads they had fantastic caps of long-haired monkey skin set with smaller bells. I lack the ability to give the readers of *Vendsyssel Tidende* an idea of the noise they produced, these thousands of people with bells, large bells, drums and horns.

They stopped outside our house. Sudden silence over the entire line as they stared at us with their green, yellow, red, blue, white and violet paint-

[3] Admiral Deinhard was Commander of the German Indian Ocean Squadron and resided in Zanzibar. He was more sympathetic to the Omani Arabs than DOAG.

ed eyes. We stood behind the window posts, each with his repeater rifle ready for shooting and ready to sell our lives as dearly as possible.

[Letter printed in VT] 22nd December 1888

The War in East Africa
From *Vendsyssel Tidende's* correspondent; The Uprising in Pangani
The Black army outside; we under cover behind the house's pillars—that is how it stood for a moment. But, as soon as they saw us, they began to howl and scream as if possessed, as they aimed at us with guns, swung their spears and pulled back their bows, aiming many hundreds of arrows at us at one time. But a couple of Arabs brought them to order, and the mob moved so close to the house that they stood around our flag poles, where the German and the Sultan's flag flew side by side [see chapter two]. They were both taken down. The German flag was torn into a thousand pieces, while the Sultan's flag was carefully folded. Also the one pole was cut up for fire, while the other was loaded onto the shoulders of fifty blacks, who, followed by the entire mob, marched away after performing their heroic deed.

In the afternoon the flag pole was planted in the ground, and, when the Sultan's flag was to be raised, a deputation of Arabs and Blacks came with a friendly request to give them our cannon, since they wanted to fire a salute with it when the flag was raised. We had a quite new Krupp's cannon with about 100 shells and casings. It stood at the entrance to the house, and the Blacks always gave it a nasty look because, except for evil spirits, there is nothing they fear more than a cannon. We did not doubt that there was trickery behind this request, as with so much of their shows of loyalty. In short we refused to give up the weapon, because it would be signing our death warrant. The Arabs exploded in fury when their battle trick was discovered. They hissed, along with strong gestures, reminding us that, even if we had hundreds of cannons, it would not help. When they all attacked us at once, we would be killed. To that we answered, in as icy tones as we could manage, that we would, however, have the pleasure of seeing many hundreds of Arabs and Blacks bite the dust, before we fell to the ground from their weapons. This answer did not really suit them, and they went away with much less noise than when they came.

It is called 'impressing the executioner,' and that we, despite our desperate situation, so boldly challenged them, apparently meant that we grew larger in their eyes. But, if they could have cast a glimpse into the house, where we sat together and talked, they would have soon seen what a depressed atmosphere reigned over us. The Blacks still kept watch on the house, twenty men at a time, with relief every hour. They snuck up to our door to lock it from the outside, so that we could not come out. All the Blacks who previously had been on our side had failed us. We were only eight Europeans against these 3000 savages, led by the sly Arabs. Our bread had run out. We had no water to wash ourselves in, a doubly necessary refreshment here. We had only a little beer to drink, and, for solid food, a little bit of canned meat. Those were all our provisions. The only comfort we had was that our appetite had nearly left us completely. Furthermore none of us had closed an eye for three days and nights. Our tormentors, the Blacks watching outside, constantly made such a disturbance with their bell-bedecked war costumes that it sounded like an attack every minute. First one of us, then the others, stood up and grabbed his gun, thereby getting those who were about to fall asleep on their feet once again. Our rifles were always at the side of our beds, and our revolvers were always around our waist. Such luxuries as undressing we had no knowledge of. If the noise sank, and we heard only the eternal bells as if they were sneaking closer, we felt we had to stand up to defend ourselves.

My thoughts went back to Denmark. It had always filled me with solemn joy, on a clear, frosty morning, when the earth was covered with snow, to hear the merry jingle of bells, when the rural folk drove to market. Here I heard the same jingle, but it had long-since lost the dear, homey sound. It was more like we heard our own death knell, but death's agony was still held back. Now that is over with, I often think that, if I again return to the dear, familiar land and hear the merry bell music, it will no longer have the same Christmas-like sound as before. The image from Pangani with the war-painted, murderous savages will rise all too vividly from my fantasy and ruin it for me.

In the afternoon we learned that today the savages would dance, but tomorrow we would be slaughtered. They showed us at a distance how they would carve us up and chop off our arms, legs and heads. We had no doubts that they would keep their word. We prepared one room, therefore, for defence, and brought all the ammunition we could get hold of in there, so that, in the end, we could retreat there. Since it had only one door, it would be easy to defend for a time. The worst of it was that the stress of the last days had put half of our fighting force on the sick list with fever. We took care of them faithfully and cured the fever, but they were so deathly exhausted they could hardly help during a fight against the advancing savages.

The next morning [7th September] the savages again held a council on the beach. Now it was about us. But, at the same time, a warship came steaming into the coast. The joy that possessed us, when we saw it set course toward the centre of Pangani, knew no limits. We stayed ready to fight, because we knew well that the Blacks, when they had to give way for the Marines, would flee past our house and try to force their way in and kill us before the soldiers could get there.

The ship came closer and closer. I observed it the entire time through binoculars. When I finally discovered that it was not a German, but an English warship, my heart sank again greatly. I knew that the English would not fight a battle for the Germans—as low as that sounds. A boat with crew was launched, and rowed cautiously towards shore. But, at the same time, a salvo of about 2000 guns fired. The boat, however, continued to row in without returning fire. The English officer,[4] when he came near, was asked what he wanted in Pangani, and he answered that he wanted to come in and take us out. He was told that there was a war here and that we were all prisoners and that no one was allowed to land. The best he could do was to, as fast as possible, return to the ship—and he did not hesitate to follow the advice. I followed the matter through the binoculars and saw them abandon us.

[Christian Lautherborn and the other Europeans in Pangani were rescued on September 8th by Brigadeer General Mathews, the British commander of the Sultan's 1,300 man army in Zanzibar. Mathews had served five sultans as a British Royal Navy lieutenant and in 1881, left the navy to work for the Sultan. Mathews took the beleaguered Europeans, along with five leading Arabs, back to Zanzibar. Clearly, without this rescue operation, the Europeans might have been killed. Lautherborn soon moved to work at the DOAG station in Bagamoyo. See chapter 2 for more details].

[Letter printed in VT] 31st December 1888

The First Engagement at Bagamoyo
Episode from the War on the Dark Continent
Vendsyssel Tidende's correspondent from East Africa, whom all will have recognised as the author of the letters 'From Hjørring to Zanzibar' has sent a few additional descriptions of the progress of the war:

Even though the uprising rages other places, all was calm in Bagamoyo at the time after we had escaped from Pangani. But an attack was expected. However we were well supplied with Black soldiers and with all kinds of weapons, and the [German] house looked like a fortress, so solidly was it built. There was, therefore, good reason to believe that we could hold this place, and all who wished to come defend it against the rebels were encouraged to do so. At the time I had been in Zanzibar for fourteen days, and, since I longed to start working and could not forget my beautiful cotton fields the villains had ruined for me, I went to Bagamoyo.

After I had been there a couple of days, we received a message that the rebels were on the march. At eight in the morning on 22nd September 1888, the news that war against the Whites would break out spread through the town like a grass fire. We saw large groups of people who tried to hide

[4] The British Vice Consul Berkeley was sent by C.B. Eaun-Smith (the British Consul-General in Zanzibar) to check on the Indian merchants (many of whom were citizens of British India) and perhaps to rescue the Germans. It is not clear if he was in the boat that tried to land (Glassman 1995: 223).

themselves in nearby houses. At first they came up and pointed at us with their guns, rather like when boys in Denmark, during a war between schools, challenge each other with shouts and threats. But we calmly allowed them to do as they wished. We did not want war and would only defend ourselves. We had two cannons—one in front of the house. The other stood on the flat roof, where, if it came to that, we could spray the entire area with shells.

At nine in the morning, they began to fire at us. They jumped out from the houses, aimed and shot at us, and then jumped back to reload. They did this so quickly and sharply that we hardly had time to aim at them. Furthermore we were in danger of being picked off, as we were relatively few. In short we were in a serious fight. We sent a couple of well-aimed shells into the covering houses. The roofs quickly ignited. The fire spread from one house to the other and chased the attackers from their cover. So came our turn. Shells and bullets swept down between them and mowed them down like straw. I had an excellent Martini Henry rifle.[5] And, as I stood in a well-covered position, took my time and did not fire before I had an Arab in my sights. The Arabs are all leaders, and the Negroes, who are actually their victims, follow them without the right to know why they are at war.[6]

After we had fought for approximately two hours, men from the German warship *Leipzig* landed. Now the rebels were caught in a crossfire, which they only stood up to for a couple of minutes. They then fled through Bagamoyo, followed quickly by the soldiers, who shot at all who came in range and carried a gun. Since the Blacks run like deer, and the burning African sun quickly took the wind out of the fat Marines, the chase was soon over. The result was about 150 Arabs killed. How many Blacks fell I do not know, but it was obvious that we all had had the same idea; hit the Arabs as the actual leaders and let the Negroes run.

A huge number of spears and guns were brought home as war booty, and, since we expected a new attack, we barricaded our windows, so that the bullets could not fly so freely around the house.

[Letter printed in VT] 4th January 1889
(continued)

On Campaign in the Dark Continent
Picture of the War in East Africa
The rebels did not attack us, but moved a couple of miles into the interior,

---

[5] The M/71 breech-loading rifle was commonly used in East Africa at this time. It had been abandoned in Europe because the puff of smoke on firing gave away the position of the soldiers (Mann 2002: 51-53).

[6] The Germans tended to blame the Arabs for the rebellion, with little regard to differentiating various groups within the Arab and the Shirazi population. The Africans on the coast and in the interior were generally seen as victims of Arab leaders. See chapter two.

where they robbed and plundered. When we heard that there was a little town serving as their camp, we decided to drive them out.

On a beautiful, moonlit night [some time in October] we left Bagamoyo, fifteen Europeans and about thirty Black soldiers. It was about three in the morning when we arrived. We cautiously crept closer and closer and finally reached the first house. A man was hidden under the roof, and, as soon as he saw us, he fired. Naturally we answered immediately, and the entire village was immediately alive. It was a strange sight seeing the dark figures run in the white moonlight. Quickly a group faced us, ready to attack, but we drove them back with well-directed fire.

A couple of times they tried to fight through our ranks, but they were thrown back with large losses each time. Now we marched into the village, which we completely burned, since we did not want a nest of rebels so close to us. We forced the people to flee. Our Black soldiers, according to the country's custom, brought home as war booty large amounts of guns, spears, drums, along with chickens and ducks—enough to supply an entire Saturday market back home.

We took it easy for a week. There were only a couple of very interesting incidents. In our own headquarters we caught one of the worst of the leaders. He had come into town accompanied only by a slave. As soon as the rumour reached us, we rushed into town. We managed to catch the man the peopled called the blood-dripping name of 'the Butcher'[7] because of all the murders he had committed. He was imprisoned to be later shot, but first we had to have all possible help from him for our plan to catch all the leaders, whose hiding places he knew.

One of the biggest Arabs is Salem bin Abdalla.[8] He was to be our first catch, and we made plans for a large expedition. At dusk we planned to quietly leave our home and sail up the Kingani River, until we reached his estate, where we would sneak up to land the boats, surround the house, and grab him dead or alive. We got two boats from the German warship, a steamboat and a large sailboat. The steamboat would tow the sailboat and our own boat too up the river.

At 7 o'clock we boarded the boats and went towards land, but missed the river mouth. So that the Blacks would not note anything wrong and hide in the forests, we had to go back to the German warship, where we spent the day becoming familiar with this floating fort with its Krupp cannons.

---

[7] We have not been able to identify 'the Butcher'.

[8] Salem bin Abdalla was a Shirazi leader with estates along the lower Kingani River. He supported the uprising, and was sentenced to death after the uprising. However, he was spared by Wissmann and his property was returned. As he was a prominent Bagamoyo citizen, he was probably spared to avoid antagonising people of Bagamoyo (Steven Fabian (SF) fieldnotes, Tanzania, 2005; Glassmann 1995: 212–13).

The next day at four in the afternoon, we again boarded the boats. We reached the important river mouth and stayed until dark. "Don't stick your arms outside the boat! It is full of crocodiles here." This was heard from the first boat, and, sure enough, when we came a little further, we could see these monsters stick their scaly heads above the water. In the dark they looked just like a piece of old, floating wood. We remained in the river mouth until the moon came out. We then steamed up the uncountable twists and curves of the river.

After an hour we came to a place where the hippopotamuses were in such numbers that it was almost dangerous to sail by. Our steamboat huffing and puffing disturbed them a great deal. Many of them were up on the bank, no doubt to plunder some Negro's rice field, but they rushed into the water as soon as they heard us come. From the steep bank they crashed into the water like falling boulders and sent into the air such a mass of water that I thought, in this narrow river, it could be compared with an ocean wave. An accident nearly happened here. Yes, with reference to the crocodiles, this could be called an accident. One of the hippopotamuses came in under the last boat, and, if there had not been so many people in it, it would have overturned. This time they only suffered a fright. The boat was lifted up on one of the grey, powerful backs and again set in the water, with the result that all fell on each other. The steamboat continued to tow, since all seem to have recovered as much as possible. But the hippopotamuses must have become very irritated, because they pursued us for about an hour, and one could long hear their roars.

[Letter printed in VT] 11th January 1889
(continued)

On Campaign in the Dark Continent
Picture of the War in East Africa
Our two boats were pulled up close to the huffing steamboat that was towing us, and, as soon as the angry hippopotamuses' roaring died out, we held council. We agreed that, as soon as we had grabbed Salem bin Abdalla, we would sail further to Mtoni,[9] where a large band was secretly assembled. But when, during our discussion, we came to the narrowest place in the river, where both sides were tightly covered with bushes and tall grass, there began a heavy fire; and the bullets whined about our ears from all sides.

I will never forget one of the episodes here. An Arab with a pair of threatening, fanatic eyes was very calmly standing until our boat was in front of him. He then aimed his shotgun and fired a large load of shot at

[9] Mtoni was one of the two river crossings on the Kingani River where *caravans* entered Bagamoyo. The other was Winde (Rockel 2000).

us. The man next to me was wounded in the hand, and my hat was perforated and my hair torn from my head in a few places. The man then tried to flee, but he had forgotten that the bank was sunken. He could not get away as quickly as he anticipated. The result was his death. I, who thought he had wounded me and that it was serious, as soon as he had shot, grabbed my double-barreled hunting rifle, loaded with wolf-shot. I sent him both barrels into his belly, the one right after the other. He fell over and remained lying in the mud, where the crocodiles the next morning certainly found him.

On the little steamboat they had a revolving cannon, which began to reply as soon as the enemy had shot. One of the shells exploded in the middle of a tight, Black group on the bank. It must have impressed the Negroes, because we did not hear more from them.

We turned around and returned to the place where they had attacked us. We landed there. It was the first time I became acquainted with the African mud, and I found it extremely disgusting. As soon as the boat hit the bank I jumped out to make place for my comrades. I immediately sank to my knees. With all my strength I worked to get free of the bad-smelling, rotting, moving mass, but the result was that I sank deeper. It began to come up to my hips. Quickly an oar was thrown to me. I worked myself free and crawled on all fours to more solid ground.

After a difficult landing we finally all came ashore and began marching towards the little village of Kureni, whose people had attacked us. They only defended themselves for a short time. Then they ran away, and we moved into the pretty little village. I said pretty, and, if you saw it, you would agree. The houses were well-built and, with the cosy, palm leaf thatched roofs, they stood in straight rows on each side of the wide, clean street, which was planted with large, shady mango trees and decorative bushes. The walls were made of clay and stone, beautifully white-washed. They looked so fine between the green leaves that many Danish villages would envy their appearance. We brought in a large bunch of coconuts, not brown and dusty like those that hang for years at home in the grocery stores' windows, but fresh and green, so that it was refreshing to drink the almond-like milk. Caught by the friendly, peaceful picture, we stood and considered sparing the village, which had been built with so much work and order. The people themselves, however, determined their fate by shooting at us from the houses that were closest to where we sat and drank coconut milk. We were immediately on our feet and drove the rebels away. A few roofs were covered with petroleum, and, in the next moment, the town was a snapping sea of fire. The flames rose high up into the air, up into the tops of the beautiful trees full of fruit. We returned quickly to the boats, because it hurt us to see the beautiful, little village be transformed into ash.

During all of this, morning had come—a quite different performance than our North Jutland summer morning, when we, with the reins under our arms or tools on our shoulders march out to the fields. The sun peeped out over the treetops and gilded everything with its light. The cheerful apes sprang from branch to branch with wanton screams, as if they greeted the new day with pleasure. The large, white water birds swam close to the bank and looked for a fish or a fat frog for lunch. The wild doves cooed in the distance, and the colourful humming birds flew like small, shiny shooting stars from flower to flower. A long crocodile lay on a fallen tree trunk, as it watched happily with its small eyes. Yes, even the thick-skinned hippopotamus stuck its huge nose above the water to get a sniff of the wonderful, fresh morning air filled with the aroma of wild flowers and leaves, which shone with dew like clear diamond drops. All of nature breathed peace. It was only between men that war raged.

Deutsch-Ostafrikanische Gesellschaft
Bagamoyo, Sunday 24th February 1889

Dear Brother-in-law and Sister
It is about time that you hear from me again. The last time I wrote, I told you about our expedition with 'the Butcher' and our home journey from Kaole.[10] I hoped that the war would soon be over. It looked that way here. All the rebels had fled to the other side of the Kingani River, and, in Bagamoyo, the Negroes began to repair their huts, which had been badly damaged during the war. The Indian merchants who had fled to Zanzibar came back, one after the other, and everything began to return to the old patterns. We Europeans could take long walks out in the country without fearing being attacked, and we all hoped soon to be able to sing "Now Stands the Rifle Empty in the Corner, Now Are We All Men of Peace."[11]

We had, however, miscalculated to a very great degree. At the end of November there were rumours from all directions: Bushiri is in Anmars;[12] from Pangani he has two cannons and two hundred breech-loaders; he will join the rebels on the other side of the Kingani and will, with his old power, either die or win. And, as a matter of fact, on the 29th and 30th of November [1888], he sat, with his people and the rebels from Bagamoyo, on the other side of the river. A total of 6000 men set up camp about 3/4 of a mil

[10] The town of Kaole was 5 km outside Bagamoyo and known as a centre of support for the rebels (Brown 1975).

[11] This title or first line of the song was also in the original letter (in Danish) written in English, so it is most probably an English or American song.

[12] Bushiri bin Salim, a Swaheli speaking Arab born on the coast, this Shirazi plantation owner and slave trader was a charismatic leader, determined to undermine Omani and German authority in the region. His haughty, aristocratic bearing did not help him maintain the devotion of his followers (Glassman 1995: 232-236).

[3.5 English miles] from Bagamoyo. On the fourth of December the warship *Leipzig* left its anchor and set course for Zanzibar and left two, small steamboats, each equipped with a revolving cannon, so they, with their quick-firing shells, could keep everything around the house clean of the enemy's people. A revolving cannon can fire about as fast as a repeating rifle—about 30 to 80 rounds per minute.

On the night of the fourth [December 1888], Bushiri and his people occupied Bagamoyo. They dug a deep trench along the beach behind a thick thorn hedge, where it was impossible for people to penetrate. With a huge boulder they built a redoubt, so that they had such advantageous cover that, if the sailors from the warship landed, they could have shot them all without being hit themselves. It was clearly Bushiri's intention, with half his people, to occupy the beach and prevent the sailors from landing, while he, with the other half, would attack our house, force their way in and kill us. Everybody, more or less, builds castles in the sky, and many go so far as to consider them as solid buildings and become quite shocked when they all collapse at once and, as often happens, bring down the builder with them. This is how it went with Bushiri's sky castle.

On the fifth of December, in the morning, the attack began. Bushiri rubbed his hands in pleasure and gave orders to his people. It was now an easy matter to make all the Germans a head shorter, as there was no warship. The two, small steamboats were too insignificant for him to worry about. Furthermore there was no chance that we could fight off an attack from our house. So his people circled our house, but, before they had taken their places, the two, small steamboats had come quite close to land. The Blacks posted on the beach began to fire on them. Since we did not know what was happening, a company of Black soldiers with a European as leader was sent down to chase the rebels away. They were, however, received by such murderous fire that they had to seek cover in a sunken road.

Now the two, small steamboats came into service. They began to fire their cannons, as they floated ever closer to the redoubt and the thorn hedge. We followed with heavy rifle fire afterwards on the beach. As soon as the first cannon shots came from the boats, our people began to cheer and shout: "The boats have cannon." Bushiri had to move forward down to the rebels. The Germans are the best; they are the masters here. Bushiri's people had not seen much in the couple of minutes they were in the trenches and behind the redoubt, because a little more than one hundred, well-aimed shells exploded in between them and sent stone and sand around their ears, producing an incredibly large cloud of dust into which we sent our bullets as fast as we could shoot. That reception was too much, and quite unexpected by Bushiri's people. They had, after all, in Pangani, not really been in any battles, since we six Europeans would never have gone to battle with such a superior power. Neither had they fired a shot in the plundering of Dr.

Meyer's caravan,[13] and all on the way to Pangani everyone fled, when they heard that Bushiri was on the way. As a result they had been possessed by the idea that they were invincible.

The short time they spent behind the redoubt, however, caused them to think differently, and, despite all the sand they had gotten in their eyes, made them more clear-sighted. Just as when one steps in an ant hill, and the ants run between each other to get a hold of the eggs, that is how they looked when they began to flee, carrying off their dead and wounded. In the confusion nobody had command, and they soon gave up their attempt to bring others to safety and seemed all to be possessed by the same thought, to get away from there as quickly as possible. It had not gone any better for the group that was to attack the house. The two cannons had had good effect and drove the enemy back, who, pursued by our soldiers, fled with a skill that brought shame to Bushiri's noble donkeys, since, even though they exerted all their energy to catch up, they fell far back. The entire battle had taken a half hour, and Bushiri's beautiful dream was over. He awoke and found himself fleeing with all of his warriors in the lead, and, an hour after he had begun the attack, all was peaceful.[14]

But the good Bushiri played a double role. The main role was defender of the fatherland, but, when that did not work, he was a robber and slave hunter. The German's presence here would be the ruin of Africa, he explained to his people, and fired them up to a war by promising them gold and everything good when the Germans are chased off. He had been lucky in Pangani, and, since we were all gone, he had done as he wished. He was also well paid for his work, since all of our rifles and ammunition, along with our good Krupp cannon with shells, fell into his hands. Now he was supposed to work to bring back law and order, but, when he knew that a Dr. Meyer, with a richly equipped caravan, was not far away, he sent his people to plunder the caravan and bring the two Europeans in chains to his house. He got three hundred breech-loaders with that opportunity, which he distributed to his soldiers while all of the valuables Meyer had with him as gifts to the chiefs of the interior he kept for himself, along with the 20,000 rupees he received for turning over Meyer and his companion to cover the expenses of the war, as he said. The German Plantation Company had a beautifully laid out tobacco plantation,[15] which they had worked on for about

---

[13] Dr. Hans Meyer and Oscar Baumann were naturalists who were exploring the Kilimanjaro area. They were hijacked and taken to Bushiri, put in chains and forced to pay a ransom. They were freed, but the money they left with the Consul-General was never given to Bushiri.

[14] However, much of Bagamoyo had been destroyed by the December 5th attack, both from rebel fire and from wantom German carnage. About 2000 local people fled to the Holy Ghost Father's missions (see footnote 17) (Bennett 1986:157).

[15] The German Plantation Society was established by Carl Peters in 1886. It was headed by Friedrich Schröder who began work at Lewa plantation in Usambara, where Lautherborn

three years, which was a great help to the entire region as it daily employed 400 people and taught them to work, so that, for miles around Kirogive, one could see evidence of European enterprise. As a friend of the fatherland, he should have done everything to help them. But here he completely forgot his role and had the plantation destroyed and the strong tobacco plants cut down and trampled into the earth. He had his people stand watch at the house, so that no European could slip away. He would free them only after receiving a quite large sum for them (how much I have now forgotten).

When there was nothing more to steal in Pangani, Bushiri spoke to his people and told them that the Germans are still in Bagamayo, that the people there are cowardly, since, otherwise, they would have chased them away long ago. But now they would go down there and show them what kind of people live in Pangani. There is much to plunder; it is a rich land; there are beautiful women there, whom we will make our slaves. Furthermore a large caravan had arrived with an enormous amount of ivory, all of which will be our booty. There was, indeed, a large caravan in Bagamayo just arrived from the country of the *Nyamwezi*. It had a lot of large ivory; I believe about 120 to 130 pounds. We promised to protect them as long as the war lasted and provided them with good, strong brick houses and, for the storage of their ivory, a large building about 100 steps from our own, from which we could help them defend themselves.[16] We then passed out among them 100 rifles along with powder and shot to their great pleasure, because it is the highest wish of a Negro from the interior to be able to exchange bow and arrow for a rifle. The fact, however, that they did not understand how to use them soon revealed itself at the first attack from Bushiri, when they loaded them so full of powder that they blew up the barrels or tore up the stocks. Three or four of them lost a couple of fingers, when their rifle completely blew up into small pieces in their hands. After Bushiri had been fought off, they came to us to have the ruined weapons replaced with new ones and to be bandaged. The enemy had not caused them any injury, but, on the other hand, there were 20 wounded by the blown-up rifles. About 25 to 30 rifles were unusable, all of which were replaced with new ones. A couple of hours of training in loading were held, so that such accidents would not happen again.

Early on the morning of the sixth [December 1888], Bushiri had his people sneak into Bagamoyo, steal everything of value and take away sixty peo-

---

later supervised the harvesting of its coffee crop while the manager was on leave. The Society tried to support the development of tropical products such as tobacco and coffee. Schröder was known for his harsh treatment of African labour (Bennett 1986).

[16] The Germans also gave the caravan a large house near the fort to store their ivory. This was the Ratu House which later became the officer's mess. This is the building where Stanley and Emin Pasha were feted and where Pasha fell (see chapter 6). It is still standing (Fabian, fieldnotes, Tanzania, 2005).

ple as slaves, but he did not try to come near us. On the seventh he came again to steal, but, when we got word that he was in Anmars, we moved into the town and had our soldiers occupy a couple of the buildings on the street where we knew he would come. As soon as he, along with the other bandits, came slinking through the street, a couple of salvos came out of the buildings, killing ten of them. The rest ran away as fast as possible.

We were then free of them until Christmas, when they again tried an attack. It was, however, at such a distance, that it was nearly impossible to hit them. They also soon saw that they could accomplish nothing and went away. The people from Kaole had assured us the entire time that only our friends lived there, and several of the leading Arabs had visited us, until Bushiri came. Then it stopped immediately. The day before New Year's Day, early in the morning it all broke loose at once close to the house in which we sat and drank coffee. We rushed to get our guns and go to our posts. It was the boldest attack they had yet tried against us. The enemy had snuck close to the house and sought cover behind the thick palm trees, behind which they fired their rifles as quickly as possible for them. We could not cause them great injury, as long as they were behind the trees. Only now and again, when they exposed too much of themselves, could we hit them with a bullet. Our Black soldiers were forbidden to shoot, because they did not yet shoot well enough to hit within three or four inches from a distance. Only the inquartered sailors and we Europeans fired when there was an opportunity. The leader, an Arab, was closest to us, and, from his good cover, shot with, for us, an unpleasant accuracy. Then a shell pierced the tree and blew up in his stomach, so that he fell over dead. When the Negroes tried to drag him away and came out from under cover, many had to bite the grass. When their comrades tried to drag them off, they also paid with their lives. As soon as a couple of them had fallen, the others began to fall back. Since, however, they had to leave the good cove, and, since they were so close to the house, not many managed to get out of shooting range alive.

When the attack was over, we went out to bury the dead, for which we had to employ many people, since one ran constantly into a dead Arab or Negro. We had, the entire time, believed that the attack was from Bushiri, until we found a wounded Negro, who told us that they all were from Kaole, and that they had promised Bushiri to attack us and try, if possible, to chase us away. Bushiri's soldiers would no longer fight with us. They said that the Germans shot too well, and they would not go there and get killed for nothing.

From New Year's Day up until today we have not seen or heard much about Bushiri. His soldiers operate in companies of twenty to sixty men around the country, steal, capture slaves or attack us, when we have gone too far from the house with only a couple of men. For example, fourteen days ago Mr. O., with ten of our Black soldiers, had gone to the French [Holy Ghost Father's] mission.[17] On the way home they were attacked by about forty men, who tried to block his way. But the old warrior, who was in his seventies, quickly occupied an Indian mosque, where he and his soldiers found good coverage behind the thick pillars and held quite well, until we (at the house), who had gotten word where he was, sent a company of soldiers to help. They attacked the enemy in the back and drove him off.

Bushiri will, however, take any advantage of the war that he can. When he saw that fighting us was only a loss for him, he formed a new plan. He knew that the Unyamwezi caravan's ivory amounted to about 100,000 kr,[18] an amount that cannot be despised; especially for a man like Bushiri, who was only used to debts. He, therefore, determined on a night attack to get a hold of this wealth. He did not know, however, that we had distributed rifles to the men of the caravan. Otherwise he would have been more cautious. At midnight, just as the moon began to rise over the sea, he arrived with his band and tried to break down the doors to the building where the caravan was staying. At the same time he tried to climb over the high wall and had, for that purpose, brought ladders. However, the good *Nyamwezi* would not let themselves be robbed of all they owned, bravely held their position and fought like men. All those who did not have rifles used their long spears or bows and arrows, all of which were poisoned. The attackers were all Arabs, because Bushiri's Blacks had refused to take part in a battle so close to our house. The battle lasted only a short time, because, as soon as we heard the first shots and saw what was the matter, we helped as much as we could. The caravan, with its hundreds of rifles, sent such a rain of bullets into the Arabs, that they got out of there as quickly as possible.

The next morning we found five, dead Arabs, and the *Nyamwezi* danced a war dance around them, before they were buried. This defeat must have enraged Bushiri because, not only did he have to withdraw empty-handed, but approximately a fourth of the Arabs who participated in the attack died later of bullet wounds and poisoned arrows. He swore death to any *Nyamwezi* that crossed his path and to the whites and our soldiers. He has kept his word like a man, since every *Nyamwezi* that he later caught and three of our soldiers were given by him to the *Does*. The Negroes from the *Zaramo*

---

[17] The Holy Ghost Father's Mission was run by Catholic fathers from France and Alsace. They maintained cordial relations on both sides and were crowded with refugees during the rebellion. The Mission began in Zanzibar in 1863, moving to Bagamoyo in 1868. In 1877 they began to expand into the interior. Initially they worked largely with freed slaves, but later focused on working with African leaders to encourage conversion (Kurtz 1978: 72-73).

[18] 100.000 Danish kr. in 1889 amounts to app. 3.6 million Danish 2004 kr. So it is an understatement to say that this is an amount "that cannot be despised."

country are cannibals, who receive with pleasure such gifts, as we would a fat goose.[19]

All was now calm in Bagamoyo, and we could rest on our laurels. We were, however, very cautious, because one can never be sure with the treacherous Arab. The reason that everything was so quiet with us was that in Dar es Salaam, where our other [DOAG] station was, Bushiri had planned an attack together with the rebels from Kilwa about three weeks ago. He had said that the German's house in Dar es Salaam was so bad that he thought he would be able to force them out. He had not, however, considered the fine harbour there. It allows the German warship to come as close as 150 metres to the house, while, with us, where the beach is very flat, it must remain 800 metres out. The Arabs had said that the German soldiers were cowards. They chased the enemy away with their big shells, but they were not able to fight. That was given a good test, when they began their attack. In Dar es Salaam there are only five Germans with about 15 Black soldiers. They immediately signaled for landing troops, since they could not hold out against such a large group. The soldiers came ashore, and, then the Arabs, at first, would not give way, bayonets were set. It then went very quickly with a 'hurrah'. Such a fight could not last long between the strong, well-trained sailors and the thin, rice-eating Arabs, many of whom were pierced through, so that the bayonet stuck out of their backs. They soon fled with admirable speed. About 50 Arabs were killed, and many were taken prisoners.

In Dar es Salaam, a German Evangelistic Mission[20] had existed there for three or four years. There had always been peace there, while in all other areas of the coast war had raged. One of the missionaries had written a long

---

[19] The *Doe* share many similarities with the *Zaramo*, a people living inland from the coastal area of Dar es Salaam and Bagamoyo. The *Zaramo* had regular contact with urban folk and many Shirazi had Arab and *Zaramo* blood. They crowded into the towns on holidays, creating volatile urban populations and fuelling mass action (Glassmann 1995: 119-120). A military intelligence report claimed that the *Doe*, while not good in warfare, are considered cunning and secretive, inclined to surprise attacks in the bush and having some influential magicians (Naval Intelligence 1920). This may explain the tendency to see the *Doe* as cannibals, a fact we have not been able to substantiate.

[20] The German Evangelical Mission was a German Lutheran group that began work in Dar es Salaam in 1887. The Evangelical Mission initially had close connections to DOAG, which did not help its efforts to gain converts. The mission was attacked in 1888 and the missionaries fled to Zanzibar. They transferred their mission stations at Dar and in Uzaramo to the Berlin Mission in 1903. It seems that the buildings of the Berlin mission were destroyed during the uprising (Sutton 1970: 7). However, Lautherborn is discussing the Benedictine Mission of St. Ottilien Congregation, the first German Roman Catholic mission in GEA. It was established in 1887 at Pugu (a DOAG station), twelve miles outside Dar es Salaam. Arab rebels headed by Bushiri attacked the Pugu Station in January 1889, killing two Catholic brothers and one sister (Ofcansky and Yeager 1997: 30-31, Sutton 1970:1, von Sicard 1970). The ransom money may have been for missionaries from the Evangelical Mission.

article in *Deutsche Kolonial Zeitung* [*German Colonial Times*], in which he described the German mission's work as the foundation for peace, since they had worked among the natives with so much Christian love. They recommended that all of us from the company [DOAG] take their example and treat the Blacks with more love, so that the war would then end. But, when the Blacks and Arabs rebelled, the first place they attacked was the German mission, where a couple of missionaries were killed and a sister was taken prisoner by the bandits. It upset us very much to hear that the mission was destroyed and that the young sister was killed. Two other missionaries and a couple of sisters escaped while two missionaries and one sister were taken hostage by the bandits. The killing of the young sister upset us, not less because several of us had known her from Zanzibar, a pretty young girl of eighteen or nineteen. If we had been strong enough, we would have immediately revenged her death, but what can a few men do against such a force? As soon as we came away from our houses and cannons, we would come into the bush, where the Blacks crawl around with a skill that makes it impossible for us to see them, while they can shoot us down with ease. In a battle in an open field we can always make a stand, because we understand how to shoot, an art the Blacks have no idea about. For example, they always knock the sights off every rifle, since it is in their way. They took the two missionaries and a sister with them to their camp. How horrible it must have been for them to spend time with such villains. The men might be able to get through it, even though it would be far from pleasant. For a woman, however, used to living among civilized people, to suddenly be with such a gang of villains, whose Moslem faith makes them even greater villains than they were before, because it teaches them to lose all respect for women.

The Germans in Zanzibar immediately offered approximately 12,000 kr for their release, and we have had long negotiations with Bushiri through the French mission. We should have had them here already yesterday, but they were all sick and could not continue the journey; so we have to wait three days more before they might be here. We have the 12,000 kr and three Arabs to be traded for them.

Bushiri has been quiet for a long time, and several rumours were circulating (which I, by the way, believe he has spread himself to lead us into a trap) that nearly all of his people have abandoned him because they had no more to eat or drink, and because all the wells were dried up. They were all dissatisfied about the division of the war booty, because he always kept the best for himself, and only a few hundred men were left. This sounded so reasonable, that we decided to attack his camp and take him prisoner or kill him. But, to avoid falling into a trap, we first went to the French mission to see if the rumour was correct. We had a secret meeting with one of the missionaries, who told us that Bushiri had 5000 men in his camp, of which 400 had breech-loaders. He had two Krupp cannons and waited only for the moon to be full to attack us. The missionary had been in the camp and told us that, about a mile before one got there, Negroes were hidden in the bush to make sure that no one, besides the camp's men, got through.

The French missionaries are on a very good footing with Bushiri, and, when his Arabs are badly wounded, he brings them to the mission, where they are nursed and bandaged. The missionaries, however, have nothing against us. That is, we all had their good wishes that we would win, and we get good intelligence from them about the enemy's position when we want it. But we can no longer come together, as they want to remain completely neutral and are afraid that Bushiri will attack them if they act as our friends. They have asked us not to come to them anymore, because Bushiri always has spies around the mission. If we are to meet, we send a letter, and then all the secretive activity takes place in the thick bush behind the mission on the edge of the beach.

When we had learned that the rumour about Bushiri's position was false, we naturally gave up our plan immediately and were very careful at night; since we knew that he, with his cannons, would shoot into the house. Near the village of Madimola by the Kingani River we had, before the war, a station that, just like so many other stations, had to be abandoned when the war broke out. There we had a four-inch, Krupp, breech-loading cannon, which was too heavy to transport and was given to one of the chiefs in the village to store. Bushiri sent for this cannon. The chief would not give it up. He was put in chains and transported, along with the cannon, to the camp. But, we said, it is a breech-loading cannon. He cannot use it. He does have a couple of shells for it, but they are not filled with powder and cannot explode, even if he fills them. There were no fuses for them. He also does not have carding, since they were all cast into the river along with the firing pins to fire them with. We therefore calmed ourselves with the thought that the large cannon could not help him, because he could not shoot with it. He had about one hundred finished shells for the cannon from Pangani, but they could not explode either, because the fuses were not attached. The only thing that he could do was shoot a hole in the walls. We were, therefore, not especially worried. We were careful however, since we would have liked to recapture our two good cannons, if he should bring them.

The previous Saturday night at 11 o'clock, two cannon shots thundered out, one after the other. One round hit the wall, but, since it came too much from the side, it glanced off leaving only an insignificant mark. When we were at our posts, a few shells whistled by coming from where we had seen the flash from the firing. During the first few minutes we sent them, now and again, a rifle salvo, which was not answered. Everything became quiet; they had withdrawn.

Early Sunday morning we immediately set all the caravan people to work looking after the shells, while we went out to search for the tracks of the cannon. This was an easy matter since the alen-high grass was completely tramped down by the Negroes who had pulled the cannon. If we had wished, we could have followed the tracks right to Bushiri's camp. None of us, however, considered that to be an especially enjoyable Sunday's excursion, so we decided to give it up until another time. Instead we looked for the place where he had fired the cannons, which we quickly found. We could see that there were two different cannons, because there was a wide and a narrow track where they had recoiled. I recognized the small track immediately as the Pangani cannon. What, however, had caused the wide track? Could it be the large Madimola cannon?[21] We measured it, and yes, the width was correct. But it was not possible that he could use it. No, it must have been another cannon with the same width. According to the directions of the tracks he must have shot close by the corner of the wall, so we could find the ball in one of the buildings lying in the same direction.

We had seen everything we wanted to and went to the building where two shells from the little cannon were found. They were quite whole and could be used again, but nothing from the large cannon was found. We had people search in the opposite direction. One half hour later, luckily they found a large shell set fast into the wall of a building a distance from our own. It was a shell from the large, Madimola cannon. In other words he had used it, but how? We quickly realized, by using the cannon as a breech-loader. On one side of the cannon there is a large screw shaped like a brass door-handle but much larger, so that one can screw with both hands. By turning it, you can open the cannon as you pull, just as when one pulls a drawer out of a table. Through the resulting opening one first shoves the shell, and then a piece of carding is pushed in. The screw is turned, and you are ready to fire through. Through a little hole on top of the cannon there is stuck a firing pin, which is bent. A cord is bound to this. With a sharp pull the powder near the firing pin is ignited and the cannon is fired. But as Bushiri had neither firing pin nor carding, what did he do? He screwed open the cannon and filled it with powder just like a breech-loader. Next the shell goes in, but it will not go forward because there are two copper rings around the shell, which screw into the rifling of the cannon. It is this that gives the shell its rotating movement, as it travels out. But what did he do? He knocked off the two rings and pushed the shell in from the front, put a powder fuse through the vent, and the cannon went off, just like when it is used as a breech-loader. It is true that he could not shoot accurately in that manner, but that was quite unimportant to him when he could only hit our house at night. The last time he only hit once in three shots, but he hopes for more luck next time. Then, when it is dark, he will come so close that it will be impossible for him to miss.

---

[21] The Madimola cannon was the cannon from the DOAG station in Madimola, which had to be abandoned during the rebellion

For that reason we have laid two mines on his way, which, when he has gone over them, will give him a free air trip, when he goes back. Furthermore I do not believe that he will try to attack us again. He might, at night, try to hit our house with his cannons and, if possible, cause us injury. He will not, however, come many times, because his people, in the dark, step on thorns in their bare feet, or cut themselves on sharp stones and refuse to pull the cannon at night. On the other hand these Negroes will sneak around in the bushes during the day in the vicinity of the town, and one has to have his eyes open, when one goes too far from the house. We have had all the bushes cut down and burned for five hundred metres around the house, so that one can be sure not to be surprised.

The latest news is that the people in Dar es Salaam are preparing for a new attack, which will take place in the near future. From the French mission we received news yesterday that Bushiri will poison the water in our well to get rid of us in that way. We immediately built a strong gate so close to the wells that nothing can come in there. We have also made a fence around every well of steel wire with long spikes, so that it is impossible to climb over. The gate is locked every night, so I do not think he will be able to place his poison.

I do not have time now to write more, because I have to send this letter to Zanzibar today; or it will not go with the post. So goodbye for now. I am well and strong and am doing as well as circumstances allow. We have not heard anything from Stanley during this time, on the other hand the rumour is that Bwana Heri[22] is in Usagara with a large ivory caravan, and it is believed that he will the join the rebels. To what extent this is true, I am not able to say; it is not, however, unlikely. I shall, however, when he reaches Bagamayo, make myself known.

Greetings to all my friends and acquaintances in Hjørring.
A friendly greeting to you and the children from your affectionate,
Chr. Lautherborn.

I would like to send you my photograph from here, but I will not have the opportunity to be photographed before I come to Zanzibar. Instead I will send you several interesting pictures from Africa, from the various battle sites and others. One of our Black soldiers has tried his skill as a service to me. He says that my friends in the cold country have not seen me for so long, that it will please them, even though it is only a picture of me. I very

[22] Bwana Heri was a *Zigua*-Swahili speaking Shirazi *jumbe* (chief) who joined the revolt against the Germans. The *Zigua* people were based in the hinterland of Saadani (a coastal town near Bagamoyo), with close connections to the Shirazi and Arab urban populations and cultures. See chapter two.

much doubt, however, that it will be much of a pleasure to see the picture I hereby send, but you might have a laugh out of it.

Deutsch-Ostafrikanische Gesellschaft
Bagamoyo, 6th March 1889

Dear Father
Your letter of the 27th of January [1889] arrived yesterday, and, to show you how glad I am when I receive a letter from home, I am sitting down immediately to answer you. That there are difficulties and, at times, often dangers associated with helping to introduce civilization here, I have very good evidence of. But to pull out just now, when my help is required, would sadden me greatly. As you say, I have been engaged as a cotton planter and not as a soldier. It is also of my own free will that I am here, not because the art of war attracts me, but because I want to extend my help to the promotion of a good cause. The idea of fighting for a medal would never occur to me. Indeed, if only people who, with their bravery or skill, received a well deserved medal, that would be something else. However, I have had the opportunity here to see that, if one is high enough in rank, it is not so important if one, in a battle, with pale cheeks and shaking knees, hides behind thick walls or in a ditch, you can get a medal for the work others have accomplished. No, that is one of the world's humbugs, which I absolutely do not value. If I should actually receive a medal, which I really do not believe, I would store it away as a souvenir of Africa. To bear an award on my breast for all to see has, for me, no value, but to bear the satisfaction in my heart that I have, once in my life, helped to introduce something good, that is what I strive after. When one sees this beautiful country, which is at the mercy of heartless Arabs, whose only occupation is to capture Negroes and sell them, and, even though they have been here for centuries, they have never taught the country's children anything beyond their Moslem faith, which they would have been much better without, one cannot help but get involved. Should I, like a weakling, pull out, when my help is required, and I know that I am doing good by being here, even though I thereby expose myself to being hit by a bullet? No, I cannot.

Now it is a matter of holding the couple of stations that are left, so that, when the German colonial army arrives, they can find a couple of places where they can be housed and trained (we will be getting Negroes from Egypt as soldiers). It is easier to hold the stations now, perhaps without the loss of a single person's life. But, if they are first abandoned and have to be taken back again, then many must bite the grass before it happens. I will promise one thing for yours, mother's and sister's sake; I will withdraw as soon as the army arrives. If I am given leave for a half or a full year, I will come home. But, if peace is concluded in a shorter time, so that I can again take up my work, I have thought about taking a trip to China, where a couple of my friends are officers in the English army. I have long corresponded with them and have promised to come when there is an opportunity. I could make such a trip in a couple of months, which would be, however, too short a time for a trip home.

And now a little about my opinion concerning Mr. Boshart's[23] article, which I believe, from your letter, you have invested your confidence in. Mr. Boshart writes that the English, Germans and French, Blacks and Arabs are all agreed that the German East Africa Company carries the guilt for the rebellion. That the English and the Arabs are of this opinion I am aware. The English would be happy if Germany completely pulled out. They could then move in and conquer the country themselves. They would then practice the same cruelties as with the conquest of the Indian colonies, where prisoners were bound to the barrels of cannons and the cannons fired. That the Arabs would like to have the Germans out of the way, since they know now they will prevent the slave trade, without which the Arabs could not survive, is not so strange. The Blacks themselves have no opinion of their own. When the Arabs say so, he says it is so and is satisfied. That a couple of businessmen in Zanzibar, who, this year suffer loss, when they have earned many thousands in other years, put, in their irritation, the blame on the company, I can well understand, but it is not just. The inhuman treatment which he writes about I have not seen. Good evidence for this is the long time the Blacks stay with us. Every day slaves of the Arabs come running

[23] The article CL refers to must be 'The German's in East Africa' printed in *Aalborg Stiftstidende*, January 17, 1889. The article is an account of an article written by the 'Berliner Tageblatt' correspondent August Boshart, printed in 'Berliner Tageblatt' December 16, 1888. Some additional comments are added in the beginning and at end of the version printed in *Aalborg Stiftstidende*.

Boshart's article is a harsh critique of DOAG, accusing them of having caused the uprising because of their total ignorance of the complicated situation on the coast and the caravan trade economy, and because of sheer incompetence in handling affairs in general. The description in the article of the complicated balance of power between the Sultan, the Arabs on the coast etc. is not too far from descriptions in today's academic literature. Boshart also argues that the German government should now (1889) take over responsibility to save the colonial project and create law and order as a prerequisite for a flourishing trade and the colonial project.

One passage among others that may have troubled CL's father reads 'The main part of DOAG's employees pass their time in idleness on Zanzibar, where they, without too much worry about the whole situation, carouse day and night'. This was, as we know from CL's letter, at least not his experience of the situation.

Judged from CL's letter, CL's father probably sent the article to CL. CL comments on many of Boshart's points for criticism. Boshart does not, however, comment on the relationship between the Arabs and the Blacks and does not discuss the slave trade, it is CL who adds this theme to the discussion. Boshart in general shows an understanding of the interests and motives that drove the Arabs, whereas the Arabs in CL's view are the true villains. CL has either missed this point in Boshart's analysis of the political situation or, most probably, this part of the argument did not seem relevant to CL. At least he does not comment on this part of Boshart's article whereas other parts of Boshart's argument is vigorously disputed.

to us here in Bagamoyo and work for us, because their masters do not treat them well. When we were captives in Pangani, our people came every day to the house to see after us. After we had arrived in Zanzibar, it was only a few days before we had them with us again. The reason for this was, no doubt, the Arabs' inhuman treatment. Our Black soldiers' rage against the Arabs knows no limits, after they have been with us and learned to be free men who earn their wages and who know that their sisters and brothers have to work for nothing and are in danger every day of being sold and sent to Arabia. We have to take good care of the Arabs who fall in battle against us, otherwise both their arms and legs will be cut off. The reason for this is, no doubt, because the Arabs have not treated them well.

Since Mr Boshart wrote all of his remarks in Zanzibar, it is strange that he did not see the inhuman treatment the Blacks suffer under the Arabs, whose slaves they are. From the customs house in front of the Sultan's palace, all day they have to bear burdens too heavy for their strength, so that there are large, hardened growths on their shoulders resulting from the pole to which the burdens are bound and the ends of which are borne by two Negroes. For this heavy work the slave gets his food, which is often not enough. But there is no lack of clubs when the poor man gives under the burden's weight. As I have been eyewitness to these scenes, there is no mistake. I must say, however, that it is the only place where I have seen the slaves badly treated. Other places in Zanzibar, as well as on the mainland, they are, for the most part, well treated.

Mr. Boshart then writes that the rebellion has not lost any of its strength as good evidence of how deep the unrest is. He also writes that it has not been possible for the company to retake a foot of land. This clearly proves that he does not know what he is talking about. As I have written in a letter to Carlsen [CL's brother-in-law], we retook the country here and, in Bagamoyo, one has again begun to build, and business and daily life has returned to normal. If the company had not forbidden our expeditions, it is a big question whether or not Bushiri would have ever come over the Kingani River. If the German navy had been allowed to land its troops, we would have, in one battle, crushed Bushiri and his power. But the navy is only allowed to protect us, not to attack. He also writes that the Blacks are forced by us to fight to defend their homes; this shows how little he knows about conditions here. Let us consider things a little closer. Who is the headman of the rebellion and why? First is the Sultan [Khalifa], who, as Mr. Boshart correctly says, is a figurehead regent. Next is Bushiri, who is one of the greatest villains and thieves I have known, and has rebelled before in the country, and who, under penalty of death, is forbidden to come to Zanzibar. Further south there is Makanda,[24] a slave hunter, and so a couple of

[24] Makanda bin Mwinyi Mkuu was a Shirazi *jumbe* in Bagamoyo who joined the rebellion, becoming one of its most prominent members (Glassman 1995: 204).

more Arab villains paid by the Sultan to keep the rebellion going. In my letter to Carlsen I have written about the company's contract with the Sultan, which the other Arabs disapproved as it would give too much power to the Germans. They know that the Europeans do not keep slaves and will try to suppress the slave trade. This will be the same as expelling the Arabs, since it is their only occupation. As the Sultan is afraid of being deposed, he has to help his countrymen as best he can. For this he naturally selects Bushiri as his first agent, who the Blacks, in their superstition, believe is possessed by the devil, and who, out of fear, they follow blindly. He is naturally glad to take part in the rebellion, because he owes God and everyone, and, in such a war, there is always something to plunder. He fights, not as a defender of the fatherland, but robs and plunders wherever he goes, captures the Blacks and sends them in chains to Pangani, from which he later will drive them longer north and seek, without danger, to send them to Arabia, where he will earn a rich profit from his catch. Makanda does the same. He is also collecting a stock of slaves, who, after the war, will make him a rich man. That is how it is with them all, robbing and plundering in the land of the brave who are defending their homes. The Negroes fight against us without knowing why they have been armed to defend themselves against inhuman treatment. But the Arabs, who have been here for nearly three hundred years, and whom it has become a habit for them to follow, tell them the most terrible stories. These stories, in his childish simplicity, the Black believes and goes to war to free the country of the horrible monsters called Europeans. It is not, after all, only Germans who have been affected. The English in Mombassa have also been thrown out. What is the reason?

It is with conviction that I say the Arabs are the instigators of the rebellion. The Negroes were always satisfied with working for the Germans. They paid them well, and the only inhuman treatment that they may have suffered was maybe a couple over the neck, when they were too lazy. As soon as the war is over, and Arabs are forbidden to bear arms, we will see how long it is before the Negroes are on our side. Boshart also writes that we have not tried to keep the caravan way open. But how were we to have time for that? As soon as we took over customs administration, the rebellion broke out; and we could not send men into the country. A caravan has arrived from Unyamwezi. The Arabs tried to stop it with good words. They said that, as soon as they came to the Germans, they would be robbed of all their ivory. In Unyamwezi, however, the Arabs have a bad reputation, because they only go there on manhunts. Therefore the caravan people did not listen to them but continued on their way. The Arabs, to their great irritation, could not stop them, because they were too strong. All of the caravan people are devoted to us and will absolutely have one of us Whites go with them, when they travel home, and become their Sultan.

Boshart then writes that the country is barren, unfertile and unhealthy.

He must have good vision, since he, from Zanzibar, can see the mainland and judge things there. As a farmer I can say that I have rarely seen such fertile soil than here. It is well suited for tobacco, vanilla and highland coffee and cocoa. That it is unhealthy here, and many cannot take the climate, might be true. However a couple of large ditches properly placed, so that the water would not remain standing in the low places, would help greatly the cause of health and also bring order, and, as this shall be a German colony, it will naturally come with the years. He also writes that the Germans from the company in Zanzibar carouse day and night. There was not one who could, without danger, travel from his station to Zanzibar. After they came together, naturally they had, in the European-style, to drink a glass to the happy return to Zanzibar. I do not doubt that they have drunk too much, but, in such cases, one must not judge too harshly. People who live in glass-houses should be careful about casting stones. The eminent Mr. Boshart is, according to what I have heard, a quite good popper of corks, but enough about that. Most of his article is, in my opinion, incorrect, and, when three or four of the company's officials go to Berlin, I hope they will make him answer for what he has written.

Now a little news about Bushiri's last attack, which you can send to Carlsen, as it fits with the last letter I sent him. On the 3rd of March at 6 o'clock in the morning, I was awoken by a couple of cannon shots. I had been on watch during the night and slept, therefore, very deeply. I sprang out of bed, dressed quickly, grabbed my rifle and ran downstairs. When I got downstairs, there was another shot; and one could see the gun-smoke not far from our house behind an Arab mosque. A couple of balls had hit our house and remained in the wall. We agreed to launch a sally, which I can only describe as seen from the house, since I could not take part. I had to stay home with the house to make sure the enemy did not attack our people in the back. Fifteen marines, twenty-eight Negro soldiers, three of the company's men and a marine lieutenant was the entire sally force. They divided into two parties and attacked the enemy from two sides. They had gone at a run from our house to where the caravans were. The entire time they had the house where the caravan people live between them and the enemy, so they were not noticed before they were close up to them. They then fired the Pangani cannon. One of our Black soldiers was so close that the air pressure tore his cap off and gave him a blow in the face, so that he turned around three or four times.

It was the last shot Bushiri's people had time for. As soon as they saw the marines and our soldiers, they immediately turned around and took flight, dragging the cannon with them. But they were pursued by the soldiers. Several were shot down, and they finally gave up dragging the cannon any further. Instead they left it and tried to get to safety. The cannon was immediately brought home by our Blacks. The other, large Madimola cannon stood in the opposite street and was circled by a group of Arabs, who tried to defend it. But, when the little cannon was retaken, the soldiers came back to help with the other one. The Arabs came under a double fire, which they tried to get out of as fast as possible and ran away without trying to bring the cannon with them. The two stolen cannons were, in short, back in the hands of their rightful owners. Bushiri's brother, a tall Arab with a long, grey beard, was the artillery man with the large cannon and had a Mauser repeating rifle, which Bushiri had stolen from Dr. Meyer. Something must have been wrong with the rifle, or perhaps he did not understand how to use it, because he didn't fire one shot, but ran away as fast as he could. During the flight he took a bullet between his shoulders and lost his rifle. It was taken up by one of the marines. He must have, however, taken several bullets, because, a little ways further up the street, he suddenly collapsed. He was surrounded by several Arabs and carried off. He died the next day of his wounds.

A half hour after the soldiers had left the house, Bushiri and his bandits were so far away that we could not reach them. We went up into the town to find the wounded and, if possible, bring them help. We found two Arabs, who we carried home and cared for until they died after a couple of hours. Our Black soldiers hate the Arabs, and we must always take care that they do not mistreat the wounded. Here one of them had seen his chance and cut the hand off of a wounded Arab, which he proudly paraded around with on the tip of his long knife. Naturally he was severely punished for this. When he was asked why he had done it, he answered that Bushiri's Arabs had cut the head off of his brother and set it on the end of a long stake (which was probably quite true), while they danced a war dance around it. The caravan people also have no mercy for their wounded enemies. A poor boy of probably eight years had had both legs crushed at the knees and lay in a pool of blood. When one of the company's men came by, he found him surrounded by the caravan people, who wanted to kill him. He embraced both of our man's knees, after dragging himself to him, and shouted: "Oh master save me. These wild Black people will kill me." He was given to a Black soldier, who was ordered to carry him to the house. When he arrived he was completely calm. He said that he knew well that the White people were better than the Arabs and the Blacks, because they took care of their wounded enemies while the others tortured them. He told me that he was there to carry ammunition for Bushiri's brother's rifle. He had seen that his master was wounded but had, at the same time, received two bullets one in each knee and could not go. Bushiri had taken a shot in the ankle, and it was only with difficulty that he got on his donkey and rode off.

Since he was an especially quick little boy for his age, I asked him if he knew what advantage Bushiri had thought to get out of his attack. He answered that Bushiri wanted to force the caravan people out of their house with cannon balls, occupy the house and bombard us from it. As soon as he had taken the house, he would have taken all the ivory to his camp. He had brought people, who had waited outside the town, to carry off the ivory. They would come when they were signalled for. He said that Bushiri had thought of nothing else the past month other than how he could get the ivory. But it must have been a hard blow losing both of his cannons along with being so wounded himself that he had to ride away. When one thinks that there were more than three hundred men, all well armed, with two cannons and plenty of ammunition for one, and that, on our side, there were only forty-seven men, one can see what cowardly villains they really are.

On our side not one single man was wounded, while, on the other side, ten were dead that we know of and thirty-eight wounded, but there are probably many more, both dead and wounded. They keep it secret as much as possible. Just like with Bushiri, because, if the Negroes first discover that he is wounded, they might lose the belief that he is possessed by the Devil and leave him. In any case, as a consequence of his last defeat, several of his people have run away and sought refuge in the French mission. There are also about 6000 refugees there, who absolutely do not share Bushiri's view of the Germans but remain neutral, since they do not know who will be master here. But I know they will be glad if it is the Germans. From Bushiri's camp we have, up to today, no reliable intelligence. Bushiri is apparently very depressed over his defeat and has no idea how he can be Sultan on the mainland, which he wants so dearly.

Concerning the little boy, on Sunday evening I helped the doctor amputate one leg. He believed he could save the other. But the little boy lost too much blood before he was found. A little past midnight he began to be colder and colder and died quietly without pain. One can clearly see what heartless villains these Arabs are, because no right-thinking man would force an eight-year old boy to go to war. But cowardice and cruelty go hand in hand. That is the last I know about Bushiri. I believe he will soon have played out his role, and law and order will return, as soon as he has received the reward he deserves. All are tired of the war and will be happy when they hear that he is either shot or otherwise gotten out of the way.

When this letter reaches you, our new soldiers will probably be here, and I will withdraw and leave the work to them. But, whether I come home or not, I cannot yet say.

Give my greetings to all friends and acquaintances.
Your affectionate son
Chr. Lautherborn

I hereby send you a German song written by the sailors on the warship *Carola* and sent to us as proof of how pleased they are with our taking of the two cannons. There are several places that do not rhyme, but one cannot require that sailors be poets.

CHAPTER FIVE

END OF WAR IN BAGAMOYO AND STANLEY'S VISIT

[Hermann von Wissmann was sent out to head a German force to quell the rebellion and arrived in Bagamoyo in the spring of 1889. He managed to end the threat of an attack by Bushiri by May 1889 and retook Pangani in July of the same year. CL took part in both assaults. Wissmann's soldiers are billeted in Bagamoyo when the chapter opens.]

Deutsch Ostafrikanische Gesellschaft
Bagamoyo, 24th July, 1889

Dear Brother-In-Law and Sister,
It is now about time that you hear a little from me. I am willing to admit that I have been lazy lately. I have also, however, had a lot to do and no peace in the evening, after work, to write. We are, you see, living a real camp life here. Mr. Wissmann,[1] with his officers, is quartered in our house. Where we had enough space before, we now live three together in one room. It can, of course, be a lot of fun at times, but mostly unpleasant when one is always being disturbed, especially when one wants to write or take care of other work. If the tower of Babel had crashed down between us, there could not be more confusion of tongues than now. Wissmann's Black soldiers have been recruited from all four corners of Africa. There are Sudanese, Zulu, Somali, Swaheli, Zaramo[2] and three to four other tribes, whose names I do not remember. Among the White officers there are Germans, Frenchmen, Greeks, Turks, and Abysinians. Each speaks their own language, which, in most cases, is not understood by the other. Every day there occur many comical mistakes, because one uses pantomime to, if possible, make oneself a little understood. If it continues this way for long, I believe I will seek employment with Folkersen[3] when I go home, because I must have finished my education by then.

[1] Herman von Wissmann was brought out to command the German forces in 1889. An experienced traveller in Africa, he was known for his decisive and brutal methods. He regarded the Arabs as the enemy of the human race, and was more sympathetic to local Africans.

[2] Wissmann's Black soldiers included 600 unemployed Sudanese mercenaries from the slums of Cairo; 300 Zulu (actually Shangaan from Portugese East Africa [Mozambique]); 80 Swaheli-speaking Zaramo who were former DOAG askari, and 80 Somalis. There were 22 Turkish police and some Europeans (Wissmann, 31 officers, 2 doctors, 1 paymaster and 56 NCOs) (Mann 2002: 47-50, 73).

[3] Lautherborn writes Folkersen, but his name was Volkersen. Volkersen played Pierrot's

Among Wissmann's Black soldiers, the Zulus interest me the most. When they first arrived, they did not have a clue about what it meant to be a soldier or about civilization either. Their clothing was little and loose, like our first parents after the Fall. Instead of a fig leaf, however, they hung a piece of skin, cut into a thin fringe and tied about the loins. They have big holes in their ears, which they use for hanging small necessities. In one ear they have their snuff in a dried leaf. In the other they have a toothbrush, not one like the one we civilized people use, but a short, green branch of a hardwood tree, of which one end is chewed into fine threads so that it resembles a painter's brush. They clean their teeth with it after every meal. Their curly mops were braided into quite long, thin braids together with grass (somewhat like when we braid a horse's mane with hay). The hair stuck out from all sides just like the needles on a porcupine. When they put their uniforms on, it was a laugh to watch. Their uniform consisted of short pants and short-sleeved shirts. It was undoubtedly the most difficult work in their lives. It took over an hour before they learned to put on the two pieces of clothing. Some used the shirts as pants, others the pants as shirts. They then put the pants on backwards and the shirts in the same way or inside out. When they finally had their clothes on correctly, it was time for the red cap. It was nearly as difficult to get this put on as it was to put the clothes on correctly. The wooly mop, which had been, for so many years, used to sticking out on all sides, would not, within a few minutes, fall in place for a cap. The stiff braids were pulled into the sides of the head and the cap put on, but the slightest movement freed them from their captivity, and they flew up in the air. The cap then sat balancing on top of the braids and soon fell down. The next day they sat on the streets and cut the braids off of each other, so that they could get the cap on. They said that they wanted to look like the other soldiers.

Their first target shooting was also very funny. Most of them had been used to bow and arrow not rifles. While they sighted, they shook with fear over what would happen when the trigger was pulled. That was only the first day however. They now shoot no worse than the other Black soldiers. They become more civilized every day. Now, when they walk by, instead of snuff and a toothbrush, they have a cigar in one ear and a bullet in the other. They say that it takes a longer time to get a bullet from the bag than from their ear. They also say that, if Bushiri comes, they will be prepared. Furthermore they are brave and proud. They say "we are wild men and have never known about soldiering, but it will not be long before we can do as much as the others"; and they do all they can to learn and will, with time, become good soldiers.

---

part at the pantomime theatre (a dumb show) in Tivoli, Copenhagen, for almost 50 years from 1843 onwards. In the winter-season, when Tivoli was closed, he toured Denmark doing his Pierrot, so Volkersen was a well-known public figure in all of Denmark in the late 1880s.

The Sudanese remind me quite a bit of the American Negroes. One finds among them many fine fellows, but, as is usually the case with Negroes, when they live well and do not have much to do, they become uncontrollable. When the Sudanese first came here, they were all very thin, starved and half-naked. Now they all have good clothes, are well supplied with food and earn good pay every month, and they live in very good housing. This change from poverty to prosperity has been too much for them, and there must often be employed hard punishment to keep them in check. They are all old soldiers, very brave, and many of them wear medals from the war with Sudan.

The Somali Negroes are actually the bravest tribe found in Africa. Most explorers have been able to acquaint themselves with the greatest part of Africa, despite the savage tribes' attempt to block their way, but they have not yet been able to travel through the Somali's country. The few who have tried have paid with their lives. The Arabian slave hunters also prefer other places for their plundering, and, up to today, no Somali has been a slave. Even though I have been used to seeing many Negroes every day for many years, I have never found any who could measure up with a Somali in understanding. His head structure is also different from all other Negroes. The high forehead, sharp nose, thin lips and soft hair are very European. Yes, I dare say that I find facial features more noble and beautiful among many of them than among Europeans. I remember seeing in Aden a young Somali in an Arabian coffee house. I count her among the most beautiful girls I have seen in my life. The beautiful, slender figure can be much better judged here than in Europe, since the dress only consists of a piece of white cloth reaching just over the ankles and up to the arm pits and wrapped a couple of times around the body and tied in a knot over the shoulder. This leaves both arms and one shoulder completely free. Yes, such a pair of arms would any Somali count himself lucky to be able to place around his neck and look into the big, wise eyes—that is if he knows how to judge a pair of beautiful arms as well as we Europeans. Even though I am not fond of strong coffee, I went into the coffee house every day to enjoy the sight of the beautiful girl. If she had been White, with the fresh face of the English—enough about that, I have come completely away from what I was going to write. The Somalis also consider themselves much better than all other Negroes. They have, in all contexts, a sense of honour, which is very rare among the Blacks. I remember, for example, one morning a pair of them had come late to work, and I, in quite clear words, made them understand that they were to arrive on time just like the others. One of them cast his head back, looked amazed at me, and said: "You speak to us just like to the other Negroes (he meant the men from the Unyamwezi-caravan, whom I have written about before); we are Somalis and belong to a great tribe and are not a slave people like the others, but used to good treatment and wish that to be taken into consideration."

I answered that, if he wanted to be better treated than the others, he would have to earn it. I did not care that he belonged to a great tribe. When he did not do his job, he would have to accept being treated like the other Negroes. He left without a word, but I have received more loving looks than the one he sent me upon departure. Consequently, however, they both came to work on time every day, since they did not want the other Negroes to believe that they were treated the same way as them. Their religion is Moslem but not the same as the Arabs here. One can only admire them for the firmness of their faith. They could not tolerate the climate here, and every day they died one after the other and were buried by their comrades. But it had absolutely no effect on them. They said that it was God's will that they should die here. I do not believe that we Europeans would have waited here so long and seen one after the other die, without trying to leave as quickly as possible to a healthier climate. But, only after their numbers had shrunken significantly, would they let themselves be talked into leaving.

The Arabs remind me a lot of the Mexicans. I had always heard them described as a brave people, but that must be in Arabia, because here they are cowardly villains and very cruel. In an honest fight, man to man, they never take a stand. If, however, they are twenty against one, their cruelty comes out into the daylight. A couple of months ago, for example, they captured one of our workers, a Black mason, and Bushiri commanded that both of his hands should be cut off and then let him go. The poor man came to our station half dead from loss of blood, without hands and with the message from Bushiri that this was the way we would also be treated when he caught one of us. We bandaged him quickly. He is now completely well, receives a pension from which he can live well and has a pair of artificial hands from Berlin.[4]

I have had a good chance to study Arabs here and have come to the conclusion that it is your best friend that you have to watch out for most. They nearly all make use of the most fawning courtesies. If one visits them in their homes, they will often use such an expression as: "You are my lord and master. I am your large dog, your footstool," or something similar. At this time, however, it is not healthy to let the sun go down before concluding one's visit, because it is absolutely not against his religion to, in an unguarded moment, shove a knife through the ribs of his lord and master. The Koran does not forbid killing a Christian dog. The eyes are the win-

[4] The Mason, Dunia, worked for the Germans, but seems to have deserted them. Bushiri's forces captured him and upon learning that he had worked for the Germans, cut his hands off. This was a turning point in Bagamoyo because Bushiri had been negotiating a cease-fire with the Germans. Wissmann used Dunia's misfortune as an excuse to break off negotiations and move against Bushiri. The severing of Dunia's hands also buttressed German claims that the Arabs were savages that had to be stopped, and that the Europeans would have to save the Africans (Brown 1975:81; Fabian, fieldnotes, Tanzania, 2005).

dows of the soul, and he seeks to close them as much as possible, so that one cannot look in. This is because he knows, himself, how black it is in there. If you speak with him, he will stand with head lowered, or his eyes will seek an incredible number of objects. He cannot look at you directly in the eye. He will assure you of his great friendship while looking to the side. He never gives a direct answer but 'if God wills,' or 'maybe,' or 'I will try' is the most common answer. When I first came here I often wondered about the way they build their houses. In a hot climate air and light are necessary, but here most houses are without windows, so that it is very dark inside. But, since he, for the most part, has a guilty conscious, the dark house fits him fine.

The fact that Arabs have won so much support among the natives here he first has his religion to thank. It fits the Swahili nature perfectly. Nearly all Negroes are, of course, fanatics, and a religion without ceremony, like ours, is, for him, without value. The Moslem religion, when one prays, encourages a bit of French gymnastics, such as kneeling with both legs at once and striking one's forehead on the ground. Then, at certain times of the year, one must not eat the entire day. It is first in the evening after six that one can begin to eat. These things are very attractive to the Negro. He goes through all of the various rituals precisely and is happy in the belief that, when he dies, he will surely go to heaven. There is then the Arab's slowness, which is just like the Negro's. A piece of work which a European would take a day to complete, will take him at least ten to twelve days to finish. First there has to be a great council. If one does not reach agreement today, the leader will rise up and say, if God wills, tomorrow. Perhaps he will finish the next day and begin the work, but, when it is half finished, there has to be another council because of some disagreement or other. The council might take a couple of days, and then the work is begun again and so further. So it always takes an extremely long time with everything the Arab does, but it is precisely this slowness that has helped Arabs here. The Arabs first taught the coast Negroes their faith, won followers and travelled to the interior, where they captured slaves and later sold them in Arabia. That the Arab has had no other interest in Africa than to earn as much money as possible can be seen by the condition the country is in now. It must be up to four hundred years that the Arab has been here, but the country shows no signs of his enterprise besides slave trading and slave hunting. Teaching the country's children to cultivate the land or teaching them a trade has never occurred to him. He has only taught them his faith, and that only to be able to control them.

Finally it cannot be said that the Arabs are liked by the natives. They have been used to them for a long time. Europeans are still something new here. Anything that can be called new takes the Negroes many years to become familiar with. In their language we are called *Mzungu*. This does not only mean European but also 'the incomprehensible.' The Negroes absolutely do not understand us. They often say "what does he really want here, he doesn't trade in slaves, he doesn't keep a harem, he is not Moslem, he does not work like us, but everything has to go quickly, and, when it doesn't, he becomes angry, but why? What he doesn't finish today, he can always finish tomorrow. Why so much worrying about one day?! After all we live so many days that we cannot count them. And why work so much? When we die we cannot take it with us. Planting for our children is stupid. It is much better that they work just like us. They then become much better people. When a person has enough to live without working, he usually becomes a bad person. No, Europeans are strange people. But they are clever, much cleverer than Arabs are, that is certain." That is the Negroes' idea about us.

But I can say that Blacks have begun to abandon the Arabs and go over to our side at an increasing rate. Within a couple of years, if we remain, the Arabs will have completely lost their influence here on the coast. If the natives were not so lazy and apathetic, one could make much more progress. Take it easy, however, is the saying here or *polepole*, as the natives say, that is the best way to go. I can well understand that the Negroes are lazy and slow. Why should he work? His house doesn't cost him anything. The material for it grows everywhere. From palm leaves or grass he makes a roof so tight that the strongest rain cannot penetrate it. The walls are made of clay, which is plentiful here. In three days, with four to five hours of work per day, the house is finished. It does not, of course, look like the houses at home. It is four to five metres square and two metres high to the top of the roof. When he goes in, he must always bend over and can only stand upright in the middle of the house. When one asks why he doesn't build a bigger house, he answers, 'It takes so much work, and why should I build a big house? When I am inside, I am either eating or sleeping. If I am sleeping, I am lying down. If I am eating, I am sitting down. For that the house is big enough. When I want to walk, I go outside. If I want to have a long talk with my friend, we sit down under a tree. Why should I do such unnecessary work?'

His diet consists of different types of roots and greens, which grow here in large amounts. When he is living really well, he eats dried shark with rice. The most important food, which is consumed over nearly all of Africa, is *muhogo* [cassava], the root of a bush, which is used in many different ways. It is eaten raw, or it is laid in the fire a couple of minutes, until it is thoroughly cooked. It is dried, pounded into meal and baked into bread. It is just as indispensable as rye for us back home. The only thing he needs to do is to stick a couple of branches from the bush into the ground each year in the rainy season, and he will have enough roots from it to feed himself and his family the entire year. His clothing consists of a piece of cloth three alen [yards] long and an alen [yards] wide bound around the hips. Hats and boots are luxuries for him. When a Negro is old enough to marry, he builds his house. When it is finished, he invites perhaps three or four of the country's daughters to be his brides, plants a field with *muhogo* and rice, digs a hole close to the house for rain water (digging a well is too much work), and works maybe an hour every third or fourth day. The rest of the time is used to eat and sleep, and to live more satisfied and happy than any millionaire. He cannot understand that others will have more, when it means work. Rather less to eat and little work than more and more work is his motto.

But now a little bit about Bushiri. When I last wrote we had recaptured the two, stolen cannons, and he had withdrawn to his camp and was quiet. Now and then he would steal a couple of Negroes and sell them. When, however, Wissmann's soldiers had trained for a couple of weeks, they paid a little visit to Bushiri's camp on 8th May 1889, which cost him more than a hundred of his people's lives. He also just barely got away himself. How is not known, since his donkey was captured and he is so heavy, he can only move with difficulty. It is believed that he was carried away. Our soldiers found many things of great value and about 6000 rupees, which Bushiri had received as ransom for an English missionary he had captured[5]. After everything usable was taken out, the entire camp was set on fire. One hour later a large pile of ashes was the only trace of Bushiri's camp. A couple of captured Arabs and Negroes were brought back. The next day one of the Arabs was hung, because our Black mason recognized him as the one who had cut off his hands. The rebels in Pangani also had to give up after a two hour battle, when we attacked them on 8th July 1889. They had stretched a thick rope over the Pangani River to block the boats. Commander Wissmann, however, had two flat-bottomed steamships equipped with revolving cannons, with which he steamed into the river. Naturally the rope broke when the ship sailed against it. When the shells began to fall among the rebels, it was not long before they retreated and we went ashore as victors. Now all is quiet there, and the Arabs have come back to ask for peace, which has been granted. However, we keep a sharp eye on them, because no one trusts them anymore. My cotton plantation is completely ruined; all the buildings are burned down, the animals are slaughtered, and all equipment broken up. There will be work for me, when I begin again, which I believe will be in a couple of months. Tanga has also been recaptured. There is only one station more, but it will not be long before that also belongs to us.

When Bushiri was driven from his camp, and there was no chance for

---

[5] The English missionaries may have been from Magila, where the Universities Mission to Central Africa had a station, or the Church Missionary Society station at Mpwapwa, on the road to Uganda (Illife 1979: 84; Glassman 1995: 245). Roscoe and his wife, from the Church Missionary Society, were captured by Bushiri and ransomed by the British. Mrs. Roscoe gave birth the day they reached safety (Beidelman 1982:57).

him here, he went with about forty men into the interior. There was a German station in Mpwapwa about 30 days journey from the coast among the *Gogo* people. Here he would go in order to revenge, if possible, the defeat he had suffered here [Bagamoyo]. At the station there was a Dane and a German, who had good relations with the natives. They had been there during the entire rebellion without noting any trace of the war that raged on the coast. They had, therefore, neither fortified their houses nor taken any protective measures. They had only passed out rifles to the people living close by and ordered that, if the station was attacked, they were all to come help. On the 8th of June Bushiri and his men (how many is not known) reached Mpwapwa and kept hidden until night, when they attacked the station. The German was awoken by a noise in the adjoining room and sprang up to see what the matter was. There was great confusion; chairs and tables were turned over, a broken lamp lay on the floor, and the burning oil created a huge fire, which illuminated the horrible scene. In a corner of the room two Arabs were cutting the throat of the Dane. The German wanted to shoot, but, since his repeating rifle just then would not function (a round was stuck in the clip), he had to flee to save his own skin. All of his people were outside, so the bullets flew by his ears. They believed that both had been killed. In the confusion among Bushiri's people, the German ran away and climbed a nearby hill followed by his people. Here he fortified himself and waited for the break of day. He then went down to the station again, buried the Dane, distributed among the people everything he had of value and returned to the hill to spend the night. The people had, however, lost their courage and did not dare stay with him. Bushiri had said that he would cut the throats of the German and all who were with him. Only three Zanzibarians would not leave him and promised to stay and bring him to the coast.

He hid in the forest for fourteen days to heal his feet, which he had cut among the thorns, when he had to flee the first night. When his feet were well enough so that he could get his boots on, he began his march to the coast, which took seven weeks. He was, during this time, often pursued by supporters of Bushiri, and often went three to four days with nothing to eat. He has his three brave guides to thank for everything. They sneaked into the fields at night and stole rice, bananas, chickens and the like, which they cooked the next day while hiding in the forest. They could travel only at night. One night, when he was so weak from his exertions and the hunger he often had to suffer and could not go further, the Blacks sneaked into a nearby village, where Bushiri was staying, and stole his donkey. They brought it to the German, and he rode it until he reached the Kingani River. He then had to set it free, because he could not get it to go in the water and Bushiri's people were on their heels. He swam across the river with the help of the Blacks and began his way towards Bagamoyo. He, however, lost his way and wandered around without recognising roads or trails. He thought that he knew the way better than the Blacks, who, the entire

'Bagamoyo'

time, assured him that it was the wrong way. He was now so exhausted that he had just about given up hope of ever reaching his friends. He allowed the Blacks to lead and, supported on their arms, continued the way they wished without seeing or hearing anything about Bagamoyo. They now had to find a hiding place for the day in order to rest for the next night. They had eaten nothing for the last two days and did not dare to go out, since they did not know the area. The German said to his Blacks, "if we do not reach Bagamoyo tonight, I do not know what we will do to get something to eat; without food I cannot go on much further." As soon as the day was over they continued on their difficult way with empty stomachs and aching legs. After two hours of wandering the cloister bells of the French mission in Bagamoyo rang and worked like a magic wand on their desperation. They reached Bagamoyo in twenty minutes of quick march.

At the mission they were hospitably received by the brothers. One of the brothers came to us the same evening to report the happy as well as the sad news. We went to the German the next day and brought him to our station. He looked very worn out, but not as bad as I had expected. Two days of good care brought him back on his feet, so that he could go to Zanzibar, followed by his faithful Blacks, who would be well rewarded for their loyalty.

Here in Bagamoyo, where I am now, everything is peaceful. The burned down town is rising from its ashes. I have had, during this time, very pleasant work, laying out the streets and clearing building sites. We will not use anymore the African construction style, with its three feet wide streets, twisting into so many turns that one has to tie the end of a piece of string to something when one goes into the town and unwind it in order to find the way out. Bagamoyo will be laid out in a new, European style with wide, straight streets. It really looks beautiful with the palm leaf covered houses in a straight line and the wide streets, completely clean and even as a macadamized road back home. The natives looked askance at the wide streets, when we first measured them out, and thought that they were too difficult to keep clean. But when I said that each man had to sweep in front of his door, so that it would all be clean, they opened their eyes wide and said, "*Mzunga hapa kazi ingi*" (Now the Europeans are here, so we must work). But they know there is no way around it and do the work without complaining.

Now I am tired of writing. In my next letter I will write about plant growth and the animal world here. I will also write about the Arabs' slave hunting and a little more about the people's lives here. I have sent you a couple of pictures from Bagamoyo along with an article written by a famous German correspondent. Later I will send you photographs from places around here. That will have to wait awhile, since I cannot get any right now.

Best wishes to all the children
Congratulations for the new-born
Yours affectionately
Chr. Lautherborn
Thank Mrs. Jensen for the flower and the greeting
CL

[a very short paragraph where CL comments on a person, otherwise unknown, mentioned in a letter from his brother-in-law has been left out]

[Letter printed in VT] 27th December 1889

From East Africa
written for *Vendsyssel Tidende* by the author of the letters "From Hjørring to Zanzibar etc."

Bagamoyo, 1st December 1889
Last night, Captain Wissmann's steamship *Vesuv* suddenly paid a flying visit here. In half an hour the ship will leave again, therefore I must hurry if I want to send these greetings to the old country where the Christmas bells

will soon jingle over a snow-clad landscape while frost-cheered, red-chinned people will amuse themselves on the ice. The steamship will connect with the French mail boat leaving for Europe.

It is Sunday morning, at 6 o'clock. The sun is shining in through the window at me. The trees are ripe with fruit. The birds are holding a loud concert of sheer joy over the cool, dewy morning. Yes, the morning and the evening are the only times when life can unfold here. During the day everything lies dead under the dreadful heat. The sand burns through the soles of your boots, and the water is almost lukewarm in the wells.

We await [Henry Morton] Stanley[6] with high anticipation. Our soldiers have been sent out to greet him under the leadership of four Europeans. The *New York Herald* has sent a correspondent who will meet him here. He has brought food and clothing in large quantities. If he succeeds in sending the first telegram about 'the results' of Stanley's expedition, he will be a most happy hack journalist, as he will then get a fee of 40.000 Danish crowns; that makes a good day's pay. In addition, the English based at Zanzibar have sent a large quantity of food to Stanley. The type of danger has suddenly changed. Until recently we all feared that Stanley would die from hunger; now one fears that he will eat himself to death.

Here at the outmost border of civilisation, yes far beyond that border, we have noticed the competition between the important papers of the world. The *New York World* also has a correspondent here. He has, without permission, 'stolen' into the interior, in order, if possible, to be the first with news about Stanley, and to send his telegram abroad before the rag *Herald* is able to print any news.

We amuse ourselves because the eager journalists have got the route Stanley travels wrong, so the one who turns round first will be the first to meet him. In three days he will be here, and then I shall be the first to send a thorough report to the people in North Jutland. Maybe they will not get the very first news. That will be sent by telegram to the millionaire-newspapers, because a telegram is very expensive; every word costs. But my report will be the fuller and more detailed.

I look forward to see the travelling gentleman after his journey through 'the dark continent.' After all the strain, all the hardships and dangers rumours have it that he has grown totally white-haired and looks like a dotard. The relationship between him and Emin Pasha[7] has grown much less warm; this we know for sure, but we do not know the reason why. Conse-

[6] Henry Morton Stanley was a Welsh-American journalist and adventurer who 'found' David Livingstone and later Emin Pasha as well as discovering the true source of the Nile among other parts of Africa and wrote numerous books about his adventures.

[7] Emin Pasha was a German Jew who studied medicine and converted to Islam. He worked for the Egyptians and later the Sudanese governments, became caught up in the Mahdist revolts

quently, another letter ought to be ready to ship off in a short while.

I have collected some spears, arrows, arches, drums and other items to ship off whenever convenient. Merry Christmas and a Happy New Year. I am sad not to be among you and to play games in your merry company. Here the game is much more serious, but I do not give up the hope of once...

The Captain is complaining. Alas, farewell for today.

[Letter printed in VT] 25th January 1890

Letter from East Africa to *Vendsyssel Tidende*

Bagamoyo, 26th December 1889
I promised to write some more about Stanley's arrival here at Bagamoyo. Here is a little about these strange days. On 4th December at 11 a.m. the world famous travellers arrived here in Bagamoyo. We had already three days in advance prepared for their arrival by decorating their rooms and the big veranda with waving palm leaves and flags, German, English, the Star Spangled Banner and the Egyptian flag. We had also collected food enough to feed an army for a whole month. But the African appetite is something special, especially if one, like Stanley and his fellows has not had any real food to eat in two years.

Major Wissmann, mounted on horseback had set out to meet Stanley and Emin Pasha. He met them by the Kingani River. The two brave travellers quickly swung themselves in the saddles of the excellent horses now at their disposal. After greeting one another heartily and solemnly with a glass of champagne at our little station at Mtoni, they rode into Bagamoyo.

We became aware of the light-footed regiment far away. Bum!....Bum!.... Bum!.... our nine canon-shots thundered and heralded that Africa once more had been crossed, that the hero of the dark continent, its civilizer, was approaching. Shortly thereafter Wissmann escorted them into our house [Ratu House], where on the cool veranda lots of small tables were laid out with the most delicious dishes this rich part of the world has to offer.

Wissmann, Stanley and Emin Pasha sat together, the rest of us as close to them as possible, to hear what the travellers had to tell. I, who thought about this letter to *Vendsyssel Tidende,* was most attentive. But the first hour or so they were so busy—let us call it dining—that next to nothing was said. At last they had satisfied their hunger and began to tell about all the dangers they had lived through in order to reach the coast.

Then new obstacles arrived. A number of German and English warships lay anchored in the bay and the officers had hurried ashore to come and

and was rescued by Stanley. Sent inland to further German interests, he was murdered in the interior in 1892.

greet the world historic persons. Soon welcomes were heard in English and German and the tale that shortly before had taken its beginning, stopped as by itself.

After a couple of hours the cannons saluted once more. Their caravan had arrived. 600 men Stanley had with him when he left for the Congo. Here at the coast, 284 men arrived. The rest were dead, from hunger and illness and as victims of the natives poisoned arrows. Their weapons were Winchester-repeaters and Remington guns. If Tippu-Tip[8] had kept his promise to Stanley, to have 600 men ready for him at the Congo river in order to carry provisions, only a few of the men would have lost their lives, and Stanley would even have been able to carry enormous riches of elephant tusks. Now he had to let them lie because he did not have the men to carry them along.

Stanley lost most men in the terrible forests he had to cross between the Congo River and the Victoria-Nyanza Lake, which according to his opinion is bigger than Germany. Only the tiniest provisions were to be found in these forests inhabited by ugly, vile and wild pygmy-people. They do not live in huts, but will cut an open space in the dense thicket where they live together in the hundreds.

These pygmy-people are about two alen [yards] high and have comparatively big, clumsy heads. Their weapons are archers and poisoned arrows. They use them with such skill that a good rifleman would be surprised. They were a most terrible enemy in these enormous forests that Stanley took three months to cross. In the thicket under the enormous mass of trees a complete darkness governed, and through this darkness they would sometimes hear a sharp whistle and in the next moment a man lay moaning with a poisoned arrow sunk into his body. Finally, when, after all these hardships, they escaped from the forest, nearly half of the men were dead.

When Stanley met Emin in Wadelai,[9] Emin was not sure whether it was necessary for him to follow Stanley to the coast. Stanley then left one of his officers with Emin, to help him. Stanley himself went back to the forest to fetch some of his men who had either deserted him or had been unable to continue because of sheer exhaustion. This took a couple of months. When he returned to Wadelai after three months, Emin and his officers had been taken prisoners by the rebels, who meanwhile fled in a hurry as the dreaded Stanley approached. Emin then had no choice if he did not want to be left alone in the middle of Africa, so he came with Stanley.

In the evening at 7:30 we had a great dinner, with thirty-six at the table, including the English, German and Italian consul, commanders from the many

[8] The Swaheli speaking merchant warlord, Tippu-Tip (1830-1905), was the most powerful Shirazi slave and ivory trader in the East African interior in the late 19th century.

[9] Wadelei was a town in the Equatorial Province of Southern Sudan.

war-ships, Wissmann's men etc. Wissmann spoke beautifully in English. He expressed his joy at seeing the radiant eyes with which the two travellers had looked at the ocean. He knew what one feels when, after travelling for years in the interior, without knowing whether one will ever reach one's goal, you once again see the boundless blue sea, and feel the fresh, cool sea-breeze. With renewed strength then come memories about home and the dear beloved ones who sit there and wait for us. He said how glad and honoured he felt to have two such outstanding men sitting at his table. Everybody had believed during the last year that they were already dead. But Stanley's outstanding knowledge of the dark continent, as well as his resoluteness had blazed a trail, now as before, and enabled him to fulfill his promise, to bring Emin help. It had cost dearly in hardship and the loss of many men's lives. But the happier we should be that it was all over now.

Stanley replied that he had never expected a reception like this. It became obvious, listening to Wissmann's speech, that he himself was a seasoned traveller in Africa. If not he would not have known what the heart most longed for after travel to the dark interior. Wissmann had helped to lighten his burden. Deep in the heart of Africa, Wissmann had sent his men to meet me with provisions and good wine to strengthen our exhausted bodies, with letters from the dear ones at home, the dear ones that for years have lived in fear because of our disappearance. He arranged for us to send letters, as his light-footed Black runners hurried to the coast in such a short time that the letters reached the mail-boat for Europe. He is right when he says that we looked at the sea with radiant eyes.

How we felt this moment is beyond description—like singing, dancing and weeping at one and the same time. Most of all we felt like school-boys when holiday starts and they can throw their knapsack into a corner and, after duty has ended, only have to think about how one most pleasantly can pass the many hours at disposal. My job, and my men's job, has now ended. Emin Pasha is here and we are happy that this man who for 15 years has lived in the interior of Africa is now sitting sound and safe among us. I will not talk about our hardships, neither of the many men whom we left behind. It is not suitable to do so at this happy moment.

Stanley continued—we have a lot of interesting news to tell. We have discovered that the Victoria-Nyanza Lake is much bigger than anyone had ever thought. We have found strange people, small and lean, big and fat. And we know that when the Egyptian praises his Nile he has none to thank for its existence but the snow-clad mountains of Uganda.

Stanley finished by expressing his thanks to Wissmann for the gorgeous meal, much appreciated when one has lived off plants and roots for a long time. He thanked Wissmann for having freed the caravan route from robbers; it made it possible for him to take it easy, now that Wissmann had disarmed Bushiri, Makanda and Bwana Heri and thereby shown that he was the right man in the right place.

Emin Pasha also thanked us for the cordiality shown him and expressed his happiness at listening to his mother's tongue after such a long time. Happiness and joy prevailed when suddenly a whisper about an accident was heard, of which the telegraph of course already has informed you.

[Printed in VT] Letter from East Africa

27th December 1889
(continued)

The information whispered finally reached all of us. Rather late in the evening, when we all still sat and enjoyed the company, Emin Pasha, only recently rescued with such great efforts, had almost killed himself. This man who had held out at his post for fifteen years among wild beasts and men, traitors and rebels, could not endure a European festivity, where every now and then somebody proposed a toast to Emin or Stanley and we all emptied our glasses. Emin therefore went to his chamber to rest. He was unaware of the Indian way of building houses, where windows reach from the floor to the ceiling, and thought the window was a door. A tin-roof below the faint-eyed man fancied to be the floor of the terrace. When the Blacks from Bagamoyo started one of their strange dances to the sound of drums he wanted to take a closer look. He went—out the window and crashed with a shout down on the tin-roof rather far below. From there he rolled down on the stone-pavement in the yard beneath, where he lay with a crack in his skull close to one of his ears.[10]

His life is in danger. I hope he will get well again, because he is such a warm and sweet person who still would have much good to do in this world. It would be a shame if he, after having lived through such hard times and so many dangers, should lose his life because of an accident here on the coast, so close to the sea ready to carry him home.

Emin Pasha is a tall, thin man with black hair and a beard. His face has sharp features with a pair of wise, loving eyes. He is born of Jewish parents, but has converted to the Moslem faith. All he wants is to win Africa for civilisation and he takes a keen interest in the colonisation of East Africa.

Stanley is a short, broad-shouldered man. His hair is white and he has a thin, white moustache. Close to his mouth is a sharp, hard line which commands respect; one would not like to be his enemy. His face is brown as a gypsy's. His eyes are the most peculiar thing about him. They are blue, very lively and never rest in the same place. Sometimes, but not very of-

[10] Whether Pasha was drunk or just near-sighted, is still a matter of dispute (SF fieldnotes, Tanzania, 2004). CL blames his ignorance on the building layout, combined with nearsightedness.

ten, they look at you with warmth. Most of the time his gaze is sharp, disinterested and fits the sharp line by his mouth.

His men do not love him. When he touched quickly during his speech on the many left behind, his voice trembled, and he paused for a moment, as if he had a lump in his throat, before he carried on. Yes, I believe I saw a glimpse of tears in his eyes. I think it is peculiar, this sudden expression of feelings, which disappeared again just as suddenly. But maybe this just shows that he is a strong-willed person. I asked one of his men whether Stanley really was a man with a soft spot in his heart? He looked at me as if I was a fool and answered: Stanley has courage, energy and wits. But a soft spot in his heart, never. He does not care how many men will die as long as he survives. Even if he had had to offer the lives of all the European men to reach his goal, he would not have hesitated. When he has set his mind on something, he will follow his goal ruthlessly, whatever it may cost. He would rather die than return without having succeeded. En route, when we were hungry he always had enough to eat. To share as comrades would never come to his mind. He does not have what we call heart and compassion. But that kind of man always fulfils what he has set his mind on—never mind the cost.

This lack of heart is on the other hand not compatible with the trouble taken by Stanley to go back and look for wounded and sick men. I would have liked to give you an impression of the journey that started in March 1887 over on the West coast of the continent, even if only a short description, but it has not been possible to piece together a full picture during the short time they stayed here. First I talked to this person, next to another one, then one of the officers. No matter what they talked about it all sounded so fascinating, like the most moving tale of adventures, and at the same time all of it is of interest for people who want to be informed.

Think about this expedition, as big as a small army, with soldiers marching into this huge, unfriendly continent, equipped to cope with whatever will happen. Soon they have to leave a troop behind, then another. It turns out that it was impossible to carry all of the equipment through a trackless territory. Illness forces others to be left behind to an uncertain fate.

When the scouts in the distance glimpse signs of people it could mean food, drink and rest. But it could also mean open fight, or sneaky treachery, or just as bad, that many men would desert the caravan, one would follow the other for fear of the dangers Stanley will meet next. It means that Stanley and his officers had to keep a sharp eye out in order to keep their men from stealing and selling weapons, ammunition, blankets—in short all the equipment, just to get a little food. It could also mean that the expedition would have to march into an area plundered and spoiled by the Arabs; this should be some of the most gruesome scenarios one can think up.

In some places village followed village, as far as the eye could see. Happy and glad people living under a smiling heaven. The rich soil provides food for all. Then one day dark skies meet the searching gaze, and at a given signal the Arab robbers plunge in on the village, like preying birds. They look like birds of prey with their sharp eyes and hoarse screams. These scoundrels have got hold of European weapons on the coast and have learned bits of the European 'art' of war. They slaughter everything that comes their way, from warlike men to small innocent children, who run after their mothers, screaming. The mothers were towed away like cows being brought to market at home in Denmark. Before the Arabs march away they set fire to the village, where before happiness and bliss dwelled, and the area is left barren, so barren that travellers will shy away from it as if it was plague-ridden. The only thing that awaits the traveller here is hunger, sickness and misery. What hardships these transports of prisoners had to endure one can only understand when one hears how troublesome the travel has been for the strong men who followed Stanley.

[Letter printed in VT] 28th January 1890

Letter from East Africa (continued)
28th December 1889

Soon they marched bravely on, confident of victory, they pushed through every hindrance. Soon they were more like poor, hungry skeletons, plagued by diseases and hunger. Where they could follow the rivers it all went rather well. But the horrors of the forest they will never forget they say. All in all, app. 500 square mil [two and a half thousand English square miles] were covered with forest.

In some places there were several tribes that made the march unbearable by constant attacks and by the use of very refined methods, like putting little very sharp stones or arrow-points on the road, covered with leaves. The carriers trod on them in their bare feet. It caused unbearable pain and stopped the march. In other places there were no signs of human beings. The forest turned into the most gruesome wilderness. Feet by feet they had to cut their way through a thicket of tough creepers and a tangled undergrowth. It nearly always rained, because the endless forests caught the steam from the rivers and soaked it in. It was wet, murky, pitch-dark, poisoned, sometimes dead silent, sometimes filled with frightening noises. And during all of this, they were exposed to deadly attacks from savages using poisoned weapons, arrows that killed by the slightest bruise of the skin.

After a while they discovered how the poison was made. In some forlorn huts in a village they found powder of dried, red ants. In the deep forest it is cooked in oil over a fire around which the savages perform secretive ceremonies. This lineament is put on the arrowheads that will kill even the travelling houses—the elephants that appear in great herds and makes broad paths in the forests; their appearance is heralded by the earth trembling and small trees crashing, as if an earthquake was under way.

One can understand that these grown men behaved like fools when they finally came out of the forest and saw green, lush grass and fresh meat instead of the moist mushrooms etc. that had served as their main food for such a long time in the darkness of the forest. Then they found Emin, as already told in many Danish newspapers, and began, after having collected their left-over equipment, the long journey towards the coast, to be greeted as I have already written about.

On 6th December Stanley went with a German war-steamer to Zanzibar, where he began the risky sport of 'eating his way home.' Every day he takes lunch and every evening dinner. The Italian scientist Kasati, who came here with Emin, is staying with us, as he for the time does not care to enjoy life on Zanzibar. Every place he would run into Stanley, and he cannot stand him. Although of older age he is as lively as a young man. Dr. Parker is very ill, but hopes to live on. The other Englishmen who accompanied Stanley are in good health and will in a couple of days set out for Egypt and home.

Besides, I have to tell you that our dear old friend, Bushiri, now is dead and gone. He, who a little less than a year ago dreamt of becoming sultan on the coast, and who ruthlessly slaughtered our men, and whose heads he cut off and put up on poles, was caught by the natives, while he wandered about, nearly naked. He only had a blue rag tied around his waist, without his men and without his donkey, after Wissmann had killed all his men. He was carried away in chains to Pangani, where Bushiri's people nearly killed, among others, your honourable correspondent. In Pangani he suddenly turned friendly and proposed, smilingly, that he should be given a house, his mind was now set on living in peace in his old age. If so he would not harm us. Well, he got a house, three alen [yards] wide. Three of his people were hanged in Bagamoyo. They took an oath on their innocence. But our Black mason recognised one of them as the man who had cut off his hands while the others had killed some of our deserted mariners. The natives tell us they are happy that these evil men are dead and gone.

Trade is flourishing again. The surrounding fields are now again cultivated. Every week a couple of caravans arrive with ivory. The roads are cleaned of robbers. A couple of these, Makanda and Bwana Heri, are still hiding in the Usagara mountains. For the time being they have no men.

Christmas passed quietly. We people from the North could not find the true Christmas spirit among all the green bushes and the baking sun. I read the other day that the most popular newspaper is *Kjerteminde Avis.*[11] Next

[11] *Kjerteminde Avis* was a very local newspaper (in the small provincial town of Kerteminde). It had 150 subscribers in 1885. So CL is being ironic. *Vendsyssel Tidende* had 1600 sub-

to this international newspaper *Vendsyssel Tidende* ought to be mentioned. The previous shipment I handed on to a German missionary, who was on his way to the Victoria-Nyanza Lake. He reads and speaks Danish rather well and said that a couple of his friends understood Danish too. When he has read the papers he will pass them on and *Vendsyssel Tidende* will therefore soon be dispersed over half of Africa.

I have written this letter once before and placed it in the envelope. Then my little monkey wanted to play postman. He may have delivered the letter but I doubt he has delivered it in the right place, so for safety's sake I have written it once more.

---

scribers in 1890 (Jette D. Søllinge og N. Thomsen, *De danske Aviser 1634–1989* Vol. 2 p. 393 (Kjerteminde Avis) and p. 532 (Vendsyssel Tidende).

CHAPTER SIX

REBUILDING BAGAMOYO AND RETURN TO PANGANI

Bagamoyo, 27th January 1890

Dear Sister

'Im Ostafrika kannt alles anders wie man denkt gehen' [In East Africa can everything turn out otherwise than one thought], as the Germans say, and I must admit that he is right. It is now almost two years since I came here to grow cotton and started my work in Pangani. After half a year the uprising broke out and I had to think of my safety and leave. It was then over with the cotton growing, so what to do instead? There was war all over the country and soldiers were more in need than workers, so I decided to play soldier for the time being and was employed in Bagamoyo. In Bagamoyo I helped to drill the Blacks. It was not too easy as I had never been a soldier myself, but by practising I went so far that my platoon became one of the best at marching and shooting.

Everything went rather well till rumours reached us that Bushiri was in Anwars. What should we do? We were only 60 soldiers against Bushiri's 6000, so we could of course not attack him, we could not even defend our own house because it was not fortified. Now we were in a mess. Fortifications had to be build around our house, but how to accomplish that as none of the Blacks knew how to build stone walls and who should take the command?

I was appointed and from the *Nyamwezi*-caravan I picked out some of the most intelligent men. I let half of the men carry stones and the other half burn mortar and build the walls. In the beginning it all went very slow, but the Blacks learned every day and worked better and better. Then an architect arrived from Zanzibar. I was just supposed to help him until he had become acquainted with particulars regarding the work and the workers, but as we everyday got more workers, it was impossible for him to oversee work on his own. Thus I could not leave, but had to stay and help him.

We worked together until Pangani had been retaken [July 1889 by the Germans], then he left with all his workers to rebuild Pangani. There was a lot of work to do and I was ordered to provide more people and continue with the building [in Bagamoyo]. I started out with two men and worked away, while at the same time letting people know we were looking out for more workers. Each day new workers turned up until I had 50 masons and now work went fast. I have now built two barracks, a big one and a small one, and a big and a small fortress. I shall send you a picture some day of

Bagamoyo Fort c. 1900

the fortresses. Today is the emperor's birthday so we do not work; instead I have revised the list of my workers who have multiplied into 60 masons, 10 carpenters and 150 workmen. I have worked for major Wissmann the last four months, but now the company [DOAG] plans to start up its activities again, and I will be in charge of rebuilding for them. Consul Vohsen[1] came here and I was asked to draw up a plan for a factory, which I in fact succeeded doing. He [Vohsen] was very satisfied and said that I should be awarded the title 'Royal Master of Building' though without the right to be called 'Your Excellency'.

The next month I started to build the factory and I at that time thought it would take me three months. As you can imagine I have had enough to do as I, besides building, have had to see that Bagamoyo gets a European lay-out. It should not come as a surprise then that I have not had time to write very often. To tell you the truth I do not have even one hour of rest a day. I am constantly on the run under the burning sun, from one place to another to make sure work is done in the proper way.

---

[1] Vohsen was the director of DOAG for a short time around 1890 (Koponen 1994:178).

'A house I built in Bagamoyo 1890'

But work is supposed to be a blessing and I consider it to be so. With what happiness can I now stroll along the streets here, lying beautifully even, where sun and wind can reach into every corner, knowing that I have accomplished this. I love Bagamoyo and its people the way a father loves his children. I know every man, woman and child. I have seen them return to the town to build their houses, and I have marked the ground and stuck the four corner-poles into the soil. I have in total marked the ground for about 5000 houses in Bagamoyo. I have been able to do so thanks to my excellent health. During my two years in Africa I have suffered very little from illnesses even though I am out in the burning sun much more than any other European.

As stated in the beginning of my letter everything can quickly turn out otherwise than expected in East Africa. The rebuilding will take at least six months, but I think I will be able to finish my task during the next month. After that I will stop as building master as Consul Vohsen has promised me that I will get a fresh start and establish a cotton plantation again. I much prefer to grow cotton but as everything is not quite back in order here in Pangani and it is still not safe to restart a plantation I have nothing against building. Building interests me too and I would not be surprised if, when the building is done, I will be ordered to do something else again, like I had no notion of becoming building master when I left Denmark, but that is how it turned out.

You state in your letter that you would like to drag me out of Africa. I cannot say that I wish to leave. I am fine. I have become well acquainted with circumstances and speak Swahili really well. That the situation in between may be a little dangerous cannot be denied, but I do whatever is possible to protect myself. If something is going to happen, I always know in advance because many of the negroes like me and will let me know, and I will then be able to take appropriate action. The negroes have four names for me, *Bwana Uzuri, Malele, Rafiki, Bwana Fundymkubwa*. The first means the handsome master, it must be ironic as I am definitely not handsome. The second name I have got by the *Nyamwezi*, it is not a Swahili name, but it means the same as the third name, 'friend'. The fourth means the great master or artist. From this you may get an idea about what the Blacks think about me.

I do not at all wish to go to a more civilised place. I fit in very well with circumstances here and am very well pleased. I think every man has a duty to do as much good as possible in life. When one lives under more civilised circumstances it takes much more talent than I have to do good. This is not the case here where I live among half-civilised. The little knowledge I have proves useful every day, and I can accomplish something useful every day that makes me as well others happy. And when it comes to dangers they exist everywhere. Here you can be stabbed by an Arab´s knife or die of fever. At home you may die of consumption or pneumonia, if at sea one may drown, and in Texas one runs the risk of breaking a neck if you fall from a horse or some crazy bandit may shoot a bullet through your head. It will never cross my mind to leave because of that.

While I am writing this letter my little mongoose (a small animal approximately the size of a rat), which I have domesticated crawls up on the inside of my trousers' leg all the time, and when he is half way up I have to stand up and shake him down again. I caught him a week ago in a sand pit where my workers dug. He bit me in my hand, but two days later he was already tame and followed me around like a dog. If I now call out his name he will pipe as a rat and come running after me. If we meet a dog or some other big animal he will take refuge on my shoulder and sit there making a loud noise until the dog has disappeared [Only four handwritten pages of the original letter have survived until today. It sounds, however, like CL was coming to the closing paragraph—we have other letters which ends with a paragraph or two about a pet animal, before the final greetings].

Bagamoyo, 6th May 1890
[Also printed in VT] 26th June 1890, 27th June 1890, 4th, 5th and 6th August 1890

Dear Brother-in-law and Sister
It is now nearly so long since I have written that it is impossible for me to remember what I last wrote about. I believe, however, that it was about Stanley's and Emin Pasha's arrival here in Bagamoyo. Yes, much has changed since then. Stanley is in England, where he probably did not receive as friendly a reception as he had expected. He has not managed to pull a veil over all the roughness with which he treated his officers and other Europeans with him on the journey, and the English press has torn into him. Emin Pasha is now as fresh as a fish. I have been with him a great deal during his stay in Bagamoyo and have learned to think a great deal of him. He reminds me so much of father, not the facial features, but his figure is exactly the same; his walk and hand movements are the same; his way of speaking when he has something important to say, with a bit of enthusiasm, is like listening to father talk about the Bologne Woods in Paris.

He went to the interior a couple of days ago with a huge caravan and shall try, by peaceful means, to acquire as much land as possible. I think, however, that it will be especially Tabora and the country between Tanganika and Victoria Nyansa, where many Arabs live and which could be of much use to the Germans, if they (the Arabs) will put themselves under their (the Germans) protection. The trip will last two years. I have worked like a horse in order to go with them, but with no success. There was too much work here, I had to stay.

Yes! I said work. It is comical to think that I left America to establish cotton plantations here in Africa, have planted cotton for five months, and, for the rest of the two years I have been here, worked as an engineer and mason. For the last year and five months I have built many large and small buildings and fortifications, which I had never before thought possible to complete; but one can get used to anything. When I first began to bring the Blacks together in order to teach them how to build, I had very little understanding or experience myself. But, by trying to teach them, I learned too every day more and more. I am now seen as an artist in the trade. It is, by the way, very interesting and fun to build here. How many of our Danish contractors would not want to be able to work throughout the year as here? Bricks are not available, but there is coral in all possible sizes; and it is very light. A little fellow of six years can drag a stone of two fod [feet] in diameter, 800 metres without much effort. We break the stones loose with strong, steel bars on the beach at ebb tide. Donkeys with baskets on each side are loaded, and the stones are carried inland. Here small, Black boys and girls carry them to the building site. We burn the mortar from the same stone, and what a mortar it is! It is so shining-white that it can be compared with snow, and it becomes nearly as hard as cement.[2] Similar to how one lays-out garden beds, two cords are stretched out in the thickness of a wall, when one will excavate a ground for a building. When the foundation is even with the top of the ground, two strong posts are set in the ground at each corner, and, between the posts, cords are stretched out as far from each other as one wishes the walls to be thick. They are stretched vertically over each other with about one and a half fod [feet] between them.

The Blacks lay bricks according to these cords, and often they lay such a beautiful, even wall that it is truly a pleasure. It often happens, however, that one or two lay a brick so close to the cord that it bends it out, and then all the others lay bricks accordingly. The wall then becomes a foot thicker in the middle than on both ends. It is a European's responsibility to stretch out all the cords, so that they are vertical over each other and to make certain that the Blacks lay the bricks correctly between and not out and in.

The Black has a very poor eye for what is even or crooked. He could never make a square himself. If he does, one can be certain that one side will be two to three fod [feet] shorter than the other. When he stretches out his cords himself, there will always be a difference of three to six inches in the wall's thickness. He measures the width between the cords with a short cord. He might let one end be a little loose. When he comes to the other end, he tightens it. There is always a difference of a couple of tommer [inches] too far out. He now checks his plumb, which he has attached to the cord above. It is maybe hanging a couple of tommer [inches] too far out. Instead of moving the cord above back, so that it is even with the one

[2] This way of building was not CL's idea, it is an indigenous building technique.

CL never mentions women's participation in building, but they are present in this picture

below, which he has just measured, he moves the cord below in line with the plumb. As he says, it makes no difference which cord one moves, as long as they are right over each other. He also lays bricks crookedly around doors and windows. He does hold his plumb, but, if the window goes inwards, and he holds the cord above, and the plumb falls in toward the window below, he says "a cord is always even, and, when the window is even with the cord, it must also be even." Satisfied with his philosophy, he walls the window crooked. I have seen buildings here built by the Blacks themselves. They are strong enough, but there is not one single place, which one can say is even. There are windows, large and small, in all possible positions, and doors that lean so far out on top that, when one opens them, they fall with a great bang. The walls are three fod [feet] thick on the bottom and four to five fod [feet] on the top. It is his way of building. When one works with them, one must have an eye on every finger and be as careful as a smith, if accurate work is wanted.

The mortar here is prepared in the following manner: a circle, four metres on average, is drawn on the ground. Within the circle, firewood is neatly stacked up to a height of one and a half metres. On top of this, coral stone is stacked in a pyramid shape up to about 2¼ metres. When it is finished

House in Bagamoyo c. 1890

it has the same shape as a peat stack at home, only much bigger. Then the wood is set on fire, and a large grass mat is stretched around it at a distance, so that the wind will have no effect on it. Otherwise the firewood would burn too strongly on the wind side, and the stones would fall down on that side without being burned. On the other hand, if it burns evenly on all sides, the stones sink quite evenly more and more together, until it all ends on the ground as a huge, glowing sphere. One then waits about twelve hours, until it is all burned, and pours water on it. The hot stones then, quite simply, fall apart. The entire mass acquires the appearance of a large pile of meal, only it is much whiter. Such a pile is called a *tano*. It contains from 70 to 80 tons of mortar and costs about 40 kroner.

When one considers the raw stones used here, construction goes quite fast. With the help of ten men I have built houses 12 metres long, 10 metres wide and 6 metres high, with corrugated tin roof in seven days. The workday is not as long here as at home. The sun rises at six, and work begins. From 11:30 until 2:00 pm one eats and then continues to work until 5:30, when work stops. This is because the sun goes down before six, and it is dark immediately. People who live some distance from the work site have to get home before that, since they would otherwise be afraid of wild an-

Caravan outside one of the new buildings in Bagamoyo c. 1890

imals. (Well that is probably enough about building; it probably becomes tiresome after awhile).

We now have winter (or the rainy season) here. For me it is wonderfully cool, but the Blacks complain when it is only 17 degrees Réaumur. They say that in such cold weather sensible people stay in the house, otherwise one will get sick from the cold. According to an old saying there are no roses without thorns. It is certainly very pleasant for us Europeans, but it also has unpleasant aspects. The birds chirp, the flowers have their scent, and the grass stands so close and high that memories of pastures back home become quite clear—memories of Sunday afternoons when I, with a full picnic basket, often spent many pleasant hours. One sits for a moment here in the grass, observing the beautiful butterflies fluttering from blade to blade. One dreams about old Denmark and imagines one sees the little, chirping lark in the air, sending out its twittering tones. Then one suddenly is unpleasantly wakened from the dream by a bite. It was a scorpion. The first task is to stand up and step on the scorpion, and the next is to wash the spot bitten with alcohol; and the next is to resolve to never again sit in the grass.

Then, in the evening, when you go to your room to write, you must

'Bagamoyo Fort' c. 1890

Bagamoyo c. 1890

Bagamoyo c. 1890

'The hospital in Bagamoyo' c. 1890

practice writing with one hand and killing an entire population of mosquitoes with the other. Then one sees a large tarantula with long, hairy legs, each one equipped with a claw. You put an upside down plate on him, and, from the claws energetically moving beyond the edge of the plate, an idea of his rage can be formed. When you finally go to bed, a centipede crawls over you with forty-two legs on each side, each and every one so hot that it could burn a hole in dry cowhide. You have to wash once again with alcohol and decide to inspect the bed from now on before getting into it. When that is over you lie down and are smothered by mosquitoes, until all of them in the neighbourhood have crawled under your mosquito net. One then slips carefully out and ties up the net as tightly as possible, so that they remain inside. You then sleep on the floor until the morning, while philosophising about the tropics, mosquitoes and the rainy season. The last mentioned, with its coolness, brings so much unpleasantness. Yes, as I write the rain pours down from the clouds as if it were poured from buckets. It is wonderful planting weather.

Throughout Bagamoyo the streets on both sides are planted with coconut, turpentine and almond trees. It looks lovely and is a pleasure in this time, when everything grows so well. What a beautiful park the Wissmann Station Building (Bagamoyo Fort) will be, laid out with broad paths on the sides planted with pineapples, large, flowering cacti, and trees related to the fir tree and quite similar to it. We found them on a peninsula not far from Bagamoyo. And naturally coconut trees, African apple trees (That is, in any case, what I call them, since they look exactly like our apples. The taste is different but very good). The Blacks have another name for them, which, however, I cannot remember. Then there comes the stink-fruit tree. The fruits have the same shape as a large, oblong gourd and weigh up to 30 pounds. Since the branches on the tree are very thin, the fruit does not hang on the branches as on other trees, but on the trunk. Until one gets used to the smell that flows out of this fruit, when it is cut in two, it is necessary to hold your nose with one hand and eat with the other. The taste is quite delicate, somewhat like a wild strawberry only better, as if a good red wine had been poured on it. There is also a tree I call a strawberry tree (botanical name not known). The fruit is green and the same shape as a strawberry, up to 7 tommer [inches] long, an average of 4 tommer [inches]. When you cut it open, it looks like a fish and tastes like a pear. Many other fruit trees and ornamental trees are now being planted, and I must say that it has made a good impression on the previously poorly maintained Bagamoyo and its residents.

It is funny, however, on the levelled paths going through the park, to see a Negro who is not familiar with the new development. He stands there totally confused, and does not know if he dares to go through or not. He makes a big detour and asks if the paths are made to go on or are only decorative. When he is told that they are made to go on, he boldly walks through and strolls as proud as a king through the paths, stamps the ground every now and again with his bare feet to see if it is solid, shakes his head, mumbles something between his teeth. He goes home, and, with certainty, comes back the next day with a large group of other Negroes and gives a lecture about European roads and how smooth they are to walk on.

The French Mission [the Holy Ghost Fathers] owns probably 400 to 500 td. [520 to 650 acres] of land, most of which is planted with coconut trees. In the centre lie all of the buildings, a large, two-storey house with a cloister, where all the missionaries or brothers live. A little ways from there is a similar building for the nuns. In the middle is the church, which is beautifully built with a high tower, with a figure of the Virgin Mary with Jesus in her arms on top. Then there is the large school building for the Black girls, who are being raised by the nuns (the mission buys small boys and girls and now have, I believe, about 200 of each. All are raised in the Catholic faith). There are also two, very long workshops, a smithy and a carpentry shop, and a couple of smaller shops, a shoemaker and a saddle-maker. A little further lies the school building for the boys, their dormitory, their dining-hall, and a large bathhouse with all new fixtures. Further along are the stalls for cows, hogs, sheep, goats and donkeys. All the mission buildings are fenced in with a high and broad cactus hedge,[3] the best fence here in Africa. This is because it is impenetrable for people and animals.

From the mission there is a broad avenue leading down to the sea. Both sides are planted with turpentine trees [mango trees], which, with their magnificent crowns, shadow the entire avenue. At the entrance, on the beach, there is a square column with a large cross on the top. I have often spent many, pleasant hours in the avenue. When the sun stands directly overhead, so that your shadow is right under your feet, it is quite hot in most places, but here, in the shadow of the mango trees' magnificent crowns, with the soft breeze from the sea, it is so pleasantly cool that one does not become tired of strolling up and down observing the multi-coloured birds in their merry play between the leafy branches. At the entrance to the mission there is a high pedestal, with a statue of the Saviour with hands outstretched, as if he would bless all who enter. Between all the buildings there is a large garden with all the possible European and African vegetables and trees, including radishes, which are year round. It is so home-like visiting the brothers in the afternoon and enjoying a light lunch, which consists of fresh bread, freshly churned butter, radishes and a good glass of red wine. Here in Africa you lick your lips over such delicacies three days later. Cabbag-

---

[3] The cactus fence put up by the Holy Ghost Fathers (HGF) was to keep out thieves but also a means for claiming the land given to them by Sultan Majid (the second Omani sultan). The locals protested and removed most of the fence. Eventually the HGF won their argument and replaced the fence (SF field notes, Tanzania, 2004).

es, turnips, parsley, carrots, celery and cucumbers are the European vegetables that grow here. The African ones are pears, eggplants, tomatoes, various types of melons, sweet potatoes, vanilla, cloves, coffee trees, cocoa trees, different types of colour trees, orange, lemon and fig trees. The last mentioned is poorly suited for the tropics. They are small and twisted and bear but little fruit. I forgot one European tree, the mulberry tree, which grows very well here and bears very sweet fruit. Indeed, I think one could introduce silk worms here in Africa, since the leaves of the mulberry make up the larva's most important nourishment.

However, the most important tree here is the coconut tree. Nothing from it goes to waste. From the leaves the so-called *makutis* [thatching made of palm leaves] is made. The leaves are removed from the branch, and bound tightly together over a stick one metre long. In this manner you can put a roof on a house so tight that, even in the strongest tropical rain, not one drop can penetrate. The leaves make baskets, mats and various other items. The nuts serve as food. They are finely chopped and cooked together with rice. The shell is cut in half, and, in each half, a long handle is set. It serves as a spoon, drinking cup and much more. Outside the hard shell there sits an approximately two tommer [inches] thick, very tattered shell, which can make excellent rope. It has the advantage over hemp in that it can, year in and year out, be exposed to rain and sun without rotting. Now for the very best, the nut, when it is green, fully grown but still without seeds, one can, early in the morning when it is still cool, pluck them and lay them in the shadows until it is hot and one is thirsty. Then a hole is cut into the nut and the almond-sweet juice is drunk. I do not know any drink that tastes so good, that is so healthy and that extinguishes thirst so well. Since there are no grapes here, one has, of course, to make wine from the coconut tree. Up in the trunk, where the crown begins, one drills a little hole and sticks in a pipe and ties a dried, hollowed-out gourd to the pipe. The sap from the tree runs through the pipe and out into the hollowed out gourd, which is collected every morning. The sap is poured through a thick cloth and is then nearly as clear as water, tastes sweet and is quite harmless. However, when fermented, it becomes as intoxicating as eighteen percent brandy. The Blacks call this drink *tembo*, and most of them are not abstainers. Finally, when the tree dies, it serves as fuel, or the trunk is cut into pieces and sold. In Zanzibar a number of items are made from them.

Every child in the mission has to learn to read, write and calculate. The day's schedule is as follows: 5:00 am get up and bathe, 5:30 mass until 6:00—until 6:30 there is breakfast, and then the school begins, which lasts until twelve noon. In the afternoon there is work in the garden and among the coconut trees, and, even though there are about 400 children, they have to work quite hard to keep all the trees in order. The boys learn a trade, and the girls learn needlework and housekeeping. When the boys reach the age of marriage, they may seek a wife from among the mission's girls. When he has made his choice, they are married and must go to the interior to help establish new stations. The French missionaries have achieved a great deal as their students, when they marry, stay together their entire lives. This proves with what endurance and conscientiousness the missionaries work, because, among the people here, it is common to acquire a new wife about every two or three months and to chase the old one away. Every rule has its exception, and that is also true here. Some stay together for six to eight years and, perhaps, in individual cases, longer, but it is very rare.

I like to go to the mission church on Sunday morning. It is wonderful to hear the children sing songs for three voices to the tones of the organ and hear old Father Superior [Etienne Baur],[4] who has now been here 30 years, hold a long sermon in Kiswaheli. When the service is over, fresh-baked bread is passed out to the children, and every face radiates with health and happiness. Then I am invited to lunch; and it is not necessary to ask me twice. Soon I am sitting among the brothers and eating, with an appetite that often surprises me, huge quantities of fresh bread, freshly churned butter, the French mission's radishes, drinking fresh milk and enjoying a good conversation, which often takes a couple of hours. It is a genuine pleasure to visit the brothers. They are so reasonable, lively and hearty that they absolutely do not despise a sinner like myself, but are always hospitable and appear to enjoy it when I visit them.

I have many friends out there, but the best is Brother Oscar. Who has been to East Africa and not known him? He has been here for twenty-four years and has worked with all of his energy for the mission. By profession he is a shoemaker and has taught many of the children to make boots. Now, however, with the continuous arrival of new, younger brothers, he has laid his profession on the shelf. He now only visits all of the caravans that go into the interior to the various mission stations [Brother Oscar was in charge of organising and furnishing the caravans that deliver supplies and news to the interior missions]. He has been through half of Africa and is an enthusiastic hunter. I have often hunted with him for wild boar and hippopotamus. We have hiked through the high grass for the half of a day. Brother Oscar would tie his habit around his waist, so that he could walk better, since he was always getting stuck in the grass. When we came home, hot and thirsty, we would retire to his room. He would fetch a bottle of the oldest and best wine, covered with a thick layer of dust, 'the thicker the better'. During a lively conversation and tales of hunting adventures he had experienced in the interior the bottle would be emptied. Brother Oscar is not really a conscientious church-goer. I have often asked him, when all

[4] Etienne Baur was the Father Superior of the HGF. He was an important negotiator during the rebellion, being trusted by both sides. He was responsible for getting Bushiri to release several British missionaries that he had captured in the interior.

the others were in church, why he was not there. For an answer I would receive: "You see they are all sinners and have, in their younger days, led a bad life. Now they have to pray that their sins will be forgiven. That is not the case with me." He then smiles with his entire face over his joke.

The greatest pleasure he had, however, was on New Year's Eve. The mission had bought five new cannons for defence in case they were attacked. He had loaded all of them to shoot in the New Year. He sent me a ticket and asked me to come out to the mission in the evening, then I would see and hear something interesting. When I came, he had a satisfied smile on his face. He took my arm, showed me his new cannons, told me how much powder he had in each and said, "tonight there will be a proper bang. In the old year we have had so much gunpowder around here that it began to be unhealthy here for both soul and body. Therefore will I fire five shots of joy over that the mission, without damage, came through the war." He nodded toward the church and said, "they have no interest in such things, therefore it pleases me that you, as a rational person, are visiting me tonight." I had to stay with him until the last stroke of the bell. He then fired, as proud as a king, his five shots. They thundered through the high palms, and I felt as if both ears were full of cotton. It was three hours before I could hear properly. He slapped me on my shoulder and said, "the mission can be proud that they have a man like me, who can exchange the prayer book for a sword when it is necessary." He then smiled and asked me to come in so that we could drink a couple of glasses of the dusty to the continuation of our good friendship in the New Year.

Everything is peaceful here. Bagamoyo gets bigger and bigger and caravans come every day. Right now I am working on a warehouse for the caravans' ivory and buildings for about 1500 men, since so many caravan people are here. The work, however, goes slowly. For nearly two weeks we have not seen the sun, only rain, rain, rain. But now the worst time is over. The reason that it has taken so long for this letter is all the work and the troubled times while Emin Pasha stayed here. Now that he is gone and the rain has begun, I have time at my disposal and demonstrate here my good will.

Your affectionate Brother and Brother-in-law,
Chr Lautherborn

Bagamoyo, 3rd December 1890

Dear Carlsen!
Your letter of the 23rd of August reached me here in the middle of November. It was a pleasant surprise. We had just had a big celebration. I had completed the building of a residence for the company, and, as it is the fashion here to have a celebration when one moves in, the representative

for our company, Hr. Baron von St. Paul, invited approximately thirty gentlemen to help us inaugurate our new house. We were all sitting at the table and had finished eating, when Hr. von St. Paul[5] stood up and gave a long speech, in which he thanked me on behalf of the company for the new house, which, happily, was now finished. He expressed his happiness and satisfaction that it had all gone so quickly and that I, in a short time in Africa, had acquired so much skill as a builder, especially when I also had to train the Blacks to do such work. He believed that I would, sometime in the future, receive a rich reward for my work, since most of the gentlemen here from Bagamoyo had me to thank that they had a roof over their heads. He hoped that I would live to see Bagamoyo grow to a large, cosmopolitan city, and that I would grow with it. Everyone then shouted a "hurrah" for me, which was, however, at the time, not so pleasant for me. When Hr. von St. Paul had concluded his speech, your letter arrived. I would have liked to read it immediately, but, out of courtesy, I had to wait until the evening.

I believe that you are correct, when you say that there could be good business in different items between here and Denmark, and have, myself, thought about sending different items home. To really earn money in that manner, however, one needs more time available than I have at the moment. Furthermore, according to my contract with the company, I do not have the right to do business on my own. It is not, however, so strictly enforced, and I can, now and again, send different things home. But I cannot enter into an actual business connection before my contract expires.

I do not know anything about kola nuts.[6] I have made many inquiries in order to provide you with information, but without result. I have even, according to the article, described the tree and the fruit to the Blacks. I have even questioned one of Stanley's Black soldiers, but they all shake their heads and say that such a fruit is not found here. I think that it must grow on the West Coast. In a couple of days, however, Hr. von Wissmann will be here, and I will ask him. If he does not know, there is not much to it.

I have a couple of crates with spears, arrows and bows, along with drums and axes and several other items, which I will send to you. You can then form an idea about how such things would be received in Denmark. If you believe that there is something to earn thereby, let me know, and I will, in the caravan time, send various items home. I have shot the beautiful, multicoloured birds you describe a couple of times and have tried to clean them. They have, however, gone to pieces each time, since they had begun to rot when I reached home. When a bird here is more than one hour old, one cannot do anything with it. There are insects here in great numbers. Among

[5] Baron Walter von St. Paul-Illaire, a former district official later turned plantation director (Koponen 1994:274).

[6] Kola nuts are found in West Africa. They have caffeine in them and are widely used as a stimulant.

them are many never seen in Europe. Two collectors here do nothing but catch butterflies and other insects and send them to Europe. I believe they earn quite well by it, but how much I cannot say.

I have written to the old folks at Christmas and sent them a selection of about ninety pictures from the war here. I have asked them to send them on to you and Caroline, so you too can enjoy seeing them. They are really quite interesting and fit the descriptions I have sent you.

I came back a couple of days ago from a trip around the coast. I am again to plant cotton for the company and needed to find a good location. I was at my old plantation in Pangani. There was not a trace left showing that people had lived there, and all the Blacks in the neighbourhood were no longer there. They had either fled or had been killed during the rebellion. I decided, therefore, to re-establish the plantation there, since a piece of good earth of about a square mil [app. 56 hectares or 140 acres] bordering the coast is as good as unobtainable. I have written the company and explained what it would cost to establish the plantation along with a detailed report of conditions. When I receive an answer, I will begin. I believe that it will probably be in February. My work in Bagamoyo is now about finished. What is left another will have to finish. I need to get started with cotton and have no time to waste. I really have to take hold to be able to have anything to show for it in the year and four months I have left.

Yes, I have begun to look forward to the end of my contract, when I can take a trip home and maybe spend Christmas in old Denmark. It has now been ten years since I have eaten Christmas cake and *klejner* [Danish Christmas pastry], roast goose with red cabbage and potatoes. I have had myself photographed and wanted to send my picture with this letter, but, since it is not finished, I will have to wait a couple of days. In the near future I will write something for *Vendsyssel Tidende*, since it is a long time since the last letter. We now have a post office here in Bagamoyo and telegraph, so, you see, we are up to date. Since the post office will be closed in an hour, I have to go with my letter, so that you may have it by New Year. I wish you, Caroline and the children a merry Christmas and a happy New Year

Your affectionate brother-in-law
Chr Lautherborn
PS: You will soon have another letter from me

Pangani, 1st March 1891
[Also printed in *VT*, 31st March, 2nd, 6th and 9th April 1891]

Dear brother-in-law and sister
It has been, for a long time, my intention to write to you, but I have had so much to do, that it has been truly impossible to find time for a long letter.

As you can see, I now live in Pangani. On the 11th I received a letter from the company instructing me to immediately begin work in Pangani to establish a new cotton plantation. Consequently I travelled from Bagamoyo to Zanzibar in order to purchase equipment; this took a couple of days. I then returned to Bagamoyo to engage workers to go with me to Pangani, as I knew that very few workers were available there. When I arrived in Bagamoyo, I assembled all of my old workers, and told them that I intended to go to Pangani to start a new station and asked who wished to go with me. I had not expected that very many would wish to go, since they all had their homes and families in Bagamoyo. To my pleasant surprise, however, they all answered "me" as if with one voice. They were, however, all craftsmen and monthly salaried workers, who would receive a high wage for such work as digging out tree roots. Ordinary workers would be good enough. Therefore I said to them that I only would use about 50 men to build the new buildings. The others I could not use, because they required too high a wage. Then one of the workers stepped forward and said: "Master, you have been our father for over two years, and, when you leave, your children will go with you. When you say you cannot give us so high a wage as before, we will work for less. If you won't take us with you, we will go to you; we know where you are going." I was happy to hear that they were so willing to go, and I promised them all work and asked them to be ready to sail in a couple of days.

The next couple of days we were busy loading quite a few donkeys, equipment and tools, and, on the morning of the 15th, about seventy men and their families went onboard. Now came the hardest part for me, saying goodbye to old friends and turning my back on Bagamoyo. We were all assembled on the beach. Before me lay Bagamoyo with an air of peace and prosperity. The morning sun shone on the new, white buildings. The birds held their morning services in the trees. The cargo sleds dragged great loads of clothing from the custom house to the various merchants, and others brought one large piece of ivory after the other, which were weighed and cleared. Everywhere one looked there were new buildings under construction, which, with the workers' cheerful singing, quickly took form. I thought back two years in time, when Bagamoyo lay desolate and abandoned. On the spot where I stood, I had three times taken part in fights against Bushiri. He was dead now and had received his well-deserved punishment. Since that time Bagamoyo had begun to rebuild. The former residents had returned, one after the other, and I had gotten to know all of them, helped them to add building by building, until the old Bagamoyo was completely built, and one began to look for more space. The beach was completely packed with people, in part my friends there to bid me farewell and also friends of my workers. The fresh morning wind began to blow, and our *dhow* (African sailing vessel) made ready to set sail. I allowed my glance to glide once again over Bagamoyo and said as brief a goodbye to my friends as decently possible, since I did not want them to see how little control I had over

'Pangani'

my feelings. Immediately afterwards our *dhow* spread its large sail and took us quickly from land, while I stood in the stern and looked at Bagamoyo until the last bit disappeared.

Sailing on an African vessel is among the most unpleasant things I know of. The ship has only four to five metres of deck by the wheel. It is otherwise quite open, so that one sits in the burning sun all day. There is no room for an awning, since the ship only has one, monstrously large sail. The winds are so unsteady at this time, that the sail must be shifted several times a day. Even though I am always out in the sun, I became so burned the first day on the ship that the skin peeled off my hands and face. The next day we had a pretty good storm, and the salt water blew against my burned face causing quite a bit of pain. Consequently I slept poorly at night. In addition I had only the hard deck, and all of my workers were sick and sacrificed to the gods, which caused a great deal of noise. We floated around on the ocean for three days and nights. On the third day, an hour before sunset, we came quite close to Pangani, so that we could see the houses.

'Pangani'

Then, the wind suddenly died and we could come no further. I have to confess that I was in a very bad humour. I was in pain everywhere from lying on the hard deck. I was stiff in all of my limbs, because I could not move on the little deck. I therefore lost my temper with the captain and told him that he was an old jackass, and that, because he had not brought us into harbour before, I would have to spend another night with him, which was very unpleasant for me. Our Black captain maintained his temper and said that it was only his job to sail; providing wind was up to God. If I could provide the wind, he would be glad to sail. I saw that he was right, and I quietly accepted my unavoidable fate.

The next morning we had a good wind, and, after struggling to come clear of the cliffs, we sailed into the river. All of my workers shouted with joy over the journey being over, and I regained my good humour. After the people were brought ashore, we rushed to get the donkeys out. The poor animals had now been approximately four days without water and had a thirst that could nearly be compared to a camel's. I then had to acquire housing for all of the people, which I was able to do in a couple of hours, after which I could rest myself.

Pangani is the most beautiful place on the coast of Africa that I have yet seen. The coast rises vertically to a height of about 30 metres. At the entrance to the river, up on the cliff, there is a small fort with three cannons that completely control the river mouth. Further up the river the cliff rises steeply to a height of 150 metres, at the foot of which there is a picturesque little village called Bweni. The cliff is overgrown with trees and grass; large flocks of goats and donkeys climb the sides of the cliff; and, from the trees, one hears the monkeys scream. The chief of the village is called Mambo Sasa. A number of his people accuse him of stealing all the oxen that used to be on the cotton plantation. I very much doubt that he is completely innocent, because, when I visited him, in the course of conversation, I asked him if he could tell me who had stolen my oxen. He looked quite uncomfortable and assured me that he did not know, that he had always liked Europeans very much, and that the rebellion was very much against his will. He showed me great courtesy and tried to redirect the conversation.

It is very difficult work reaching the top of the cliff, but, when one has arrived, there is full payment for all the effort. On one side lies the great, shining ocean, which appears to stretch so far out that it, at last, comes together with the clouds. On the other side lie a considerable number of small corn plantations. The people in the fields appear, from our high point, lilliputian. Herds of cattle are seen moving in the beautiful pasture. A large water hole in the middle of the pasture, where all the cattle are watered, shines in the sun's rays like a huge diamond. It is a genuine picture of African rural life. It is only when it is far off, and everything looks so small, that one thinks involuntarily that one is standing before a panorama. Turning around, you see an even more beautiful sight—Pangani lies beneath your feet surrounded by an incredibly large palm forest. The land then begins to rise, and you see one forest covered mountain after another, which, at a certain distance, are highlighted by fine, blue air. You can see the small huts built on the mountain's side, and small, black dots moving here and there, which, through binoculars, turn out to be people cutting down trees on the mountainside. But it is not only on the cliff's top that the land looks beautiful. A sail tour on the beautiful, wide, river is also fantastic. It twists like a snake between the mountains in such large curves, that one, after a couple of hours of sailing, is often parallel with the place one started from. It often appears as if the river suddenly disappears under the foot of a mountain, but, when one reaches that spot, a sudden turn is discovered and the shining water smiles upon us on the other side, where it continues its easy-going course.

At the foot of the mountains, bordering the river, there are a considerable number of small sugar plantations, which belong to the Arabs and are

'Bweni on the other side of the Pangani river'

worked by slaves. When one has seen the large sugar plantations in America and compare them with these here, the difference is quite striking. A hoe is used for all field work, and the mill where the juice is pressed out is turned by ten, bound Negroes. I felt pity for the poor men, who had to go round and round all day long, but they sang and joked, so they must have been satisfied with their work. Sugar is not made here. The juice is boiled and shipped in tin cans to Sangilov, Pemba and Arabia. I do not believe the Arabs earn great profits from their plantations, but they are also satisfied. They have so much that they do not need to work, but lay about all day. Where there are no plantations, trees grow quite close to the bank, and their crowns jut far out over the water. In the shadows are seen beautiful, large water birds. Often a crocodile will stretch himself out to his full length under the trees and is difficult to see, since he has the same colour as the trunks. Here and there a hippopotamus sticks its head out of the water, grunting with as much pleasure and contentment as a hog at home after he has eaten himself full and is resting in the straw.

In Pangani I visited the old house, where I was once confined by the

Arabs. It still stands unchanged. Since, as you know, it does not hold the most pleasant memories for me, I did not stay long.

Between the people of Pangani and Bagamoyo I see a great difference. In Bagamoyo the Black has more trust in the Europeans and is, in most cases, open and honest. Here he is fearful and lies so that one can almost never believe what he says. Guilt for this is, borne only by the Arabs, who are about 900 [In the VT edition is added: The Arabs here constitute the upper class]. Lying and love-making are his [the Arabs'] nature, and the Blacks, who are near him every day, have taken on some of this. When I came here, I engaged an Arab by the name of Magango to purchase palm leaves to cover the new worker housing on the plantation. I felt that he used too much money, and I suspected that he cheated me. Since he was an old man, I gave him a chance to save himself. I told him that he paid too much for palm leaves. If he could not buy them cheaper, he should not buy more, because I would give the job to another. The old man struck his breast and assured me that he went to all possible lengths to get everything cheap for me, but, since so many houses in Zanzibar had burned down, all the palm leaves were sent there and that was the reason they were so expensive. I pretended to believe him, and he continued to buy for the same price. I sent an Indian to the place where the palm leaves were bound together to ask about the price, and discovered that he had bought the palm leaves for a fourth of the money I had given him; the three fourths he had pocketed himself. I sent a message asking him to come to my room. There I charged him with his guilt and had him give me the money back he had stolen. Even then he tried to lie his way out of it all by saying that the Indian was his enemy and had made up a lie to hurt him. I said that I had noticed that the Arabs prayed to God five times a day. That was probably because they all had so many sins to forgive. If they were all as great swindlers as he was and in their old age, did not improve, it wouldn't help if they called on Mohamed all of their lives; the Devil would get them when they died. The old man vibrated with anger, because an Arab, who almost always hates Christians, cannot tolerate that anybody talks about their religion.

I called another Arab and asked him if he would take some lumber I had in Pangani on his boat out to the plantation. He told me, in the most courteous manner, that it was always a pleasure when he could be of service to a European and asked me to show him where the lumber was so that he could get it the next day. He asked me to tell my people to help him. I have to confess that I did not believe a word he said and asked him if he was quite certain he would come tomorrow. He asked why I doubted his word. He had said, from the beginning, yes, and had not tried to get out of the work. I told him that he had lied very well, but that I thought that he, in the night, would sail up the river and abandon me. I told him he should tell me if he would not come, so that I could get another. He asked me to trust that he would, without doubt, come. I ordered my workers to load the lumber in the boat for the next day, but my Arab was not there the next day. He had, during the night, sailed up the river. Luckily I got another boat, but I have reported my Arab and demanded that he be punished. I have no love for the scoundrel, and I will see to it that the punishment is not too mild.

There is a great deal of work and difficulty associated with establishing a plantation here. Just think about an incredibly large field with grass and trees but nothing else. On this place there must be built a house, the field cleared, cotton sowed and later processed, and all must be finished within a space of eight months so that I can send cotton to Germany. If I manage it, I will be the first to send products from East Africa to the German market. To achieve this task I will need all available energy. It was a pathetic sight that met me at the old plantation. Where before the lush cotton stood with its red and white flowers, there now grows tall grass. All the buildings are destroyed, and everything taken away. Only the ruins remain. In the woods there is an old Portugese well, which I had improved when I first came to Africa. The Arabs had filled it with stones and branches. I sent half a score of workers there to clean it. It was exhausting work, and while they rested a moment, I came down to them. They asked me if I had built the well. To

'The fort in Pangani'

which I answered: "No it originates from the Portugese time and is about 300 years old." Now they wanted to know who the Portugese were and what had happened to them. I told them that 300 years ago there were only very few Arabs here, then it was the Portugese who ruled. Then the Arabs came with great power and conquered the land. The Portugese had to flee. Since that time the Arabs had power in the land until now, when the Germans rule. Here I attach a little evidence for how the Blacks understand time. They had attentively followed the story, and among them was a little eight year-old boy by the name of Misjumvi. When I finished talking, one of them said to me: "You have been here very long and experienced much.You must have been very little back when the Portugese lived here, probably the same size as Misjumvi." Since it was not the first time I have heard the Blacks express themselves in that manner, the question did not surprise me very much. I tried to make them understand that a man rarely lived more than 70 years, in other words there were four old peoples' ages since the Portugese were here, and I was still a young man a little over 30 years. How then, they asked, can you talk about it as if you experienced all of it? I told them that I had read it in a book; when one could read, it was not necessary to experience everything to know what happened.

After a couple of days of difficult work, we had the well repaired and immediately drew water for the donkeys, which had gone thirsty for two days. I went there in the afternoon to see if everything was in order. I found one of my men lying on the road looking through the bushes. He signalled to me that I should be quiet, and I crept up to him to see what he was looking at so intensely. Between the branches I could clearly see the well, where we had set two, large barrels for the donkeys. Around the barrels were about seventy apes quenching their thirst. I made a movement and they all disappeared like lightning between the branches. Where had the apes got water before, I asked the Black, since the well had not had water for two years. The Black said: "Apes have understanding just like people. When we have not drawn any water, they have gotten it themselves. One of the stronger apes holds fast to the wall's edge, another grabs the waist of the first, a third does the same; and they continue, until the last one has reached the water. Then all the apes run down the living rope and drink. The apes who have formed the rope are replaced, and others take their place, when they drink." This story reminded me a little of the story about the people of Mols[7] and the thirsty tree. I have heard many tales of that kind, but, even though I have lived long where there are apes, I have never seen it.

My plantation is about ¾ of a mil [5 miles] from Pangani, and I have to go out there twice a day. It is tiresome work, especially when it is terribly

[7] Mols is a Danish landscape. Molbohistorier (stories about the people of Mols) equals stories about the wise men of Gotham in English.

'Some of my workers with their overseer'

hot. But I think I will have a house finished in a couple of weeks and will then live out there. Then I can take it easy, as we say in Denmark. The Black farm workers who lived here before have all fled, and the place looks a little forlorn. The people in Pangani told me that I should not take the donkeys out there, because there were a couple of lions in the area. I have made a very careful search, but have found no trace, so I think it is a fable. It was comical though, when I went out to the plantation early one morning and had left my rifle home, since I had no more cartridges, I heard, close by the road, a sound like something falling to the ground, and, at the same time, I saw branches move in a couple of large bushes close to me. Maybe it was true about the lion, I thought; and, for a couple of seconds, I wished I had never seen Africa. But, since I know the lion is a very timid animal, especially in the daytime, I began to shout and run toward the bushes to scare him away. At that moment, a little Black girl stuck her head out between the branches, quite grey in the face from fear. I asked her why she had not said something. She said that she was plucking berries, and the branch she was on broke just as I arrived. When she saw me run toward the bush, she became very frightened because she thought I was crazy. I talked to her a couple of minutes to convince her of the opposite, gave here a couple of pesas and we continued on our way, with great satisfaction on both sides.

If I get time, I will write and tell about my progress with the cotton, since it might interest someone. Emin Pasha is on the way to the coast. There is a great deal of talk, and much has been written about Emin and Major Wissmann, but I believe that feelings are for Emin. Major Wissmann will be here in a couple of days. He has been on an expedition to Kilimanjaro and has had a couple of battles with the Maasai,[8] who have broken the peace. As retaliation several of them have been killed, and their cattle have been driven away. I have enclosed photographs of Emin and Doc Peters; perhaps they would interest you. From Bagamoyo I have sent you three chests with arrows and bows and other things from here, which I think will win your approval. And so goodbye for now and be well. The time will soon come, when I can visit you.

Yours affectionately
Chr. Lautherborn

---

[8] Maasai are a pastoral people who live in North Central Tanzania and also in Kenya. They are patrilineal and their culture centres on cattle.

Pangani, 12th April 1891

Dear Sister,
Your long, interesting letter, to my great pleasure, reached me a couple of days ago. I see that you are well and happy and that old age has not yet begun to oppress you. When I read about your and Carlsen's trip to Copenhagen, I thought it was by car. But, when you wrote about a machine that you had to pull behind you where the roads were bad, I discovered, to my great surprise, that the trip was made by bicycle. I would have never thought it of you in your old age. The next I will expect is for you and Carlsen to pay me a visit by balloon or flying machine. I would think that the trip to Fredericia must have been very interesting and especially for father, since it contains so many memories, happy as well as glad. The manor house Hindsgavl[9] I have read about before, and I can imagine that it must be an enchanting sight. Judging from your letter, father was not the only one who was close to jumping out of his skin in enthusiasm over all the wonders there were to see.

You ask me if I know that Vilhelm holds lectures for the peace cause,[10] a funny question. What do you think I do with the *Vendsyssel Tidendes* I get here? They are read up and down, and it is a pleasure to see a little, local newspaper contain so much news from abroad. It has happened a couple of times that I have told the Germans here news from Germany, which they had not yet heard. I know naturally also that Carlsen has a good friend, by the name of Wulff, a poorly thinking man who tries to impede Carlsen in the good that he wants to accomplish. I am convinced that Wulff does not believe what he himself says and that he knows that he serves a bad cause. When a man defends a good cause and is attacked, he will always give a calm and definite answer, because his conscious tells him that he is acting correctly. If, however, he defends a bad cause, in which he himself has little faith, he easily becomes infuriated. When someone contradicts him, because he himself feels that his opponent is right and, in his irritation over this, he makes gross statements and, with every opportunity, seeks to give his opponent's good work the appearance of having some kind of dark side. But, if I know my countrymen, they can see that Wulff, despite the innocent

---

[9] Hindsgavl is a sizable manor on the island of Funen, with a beautiful main building and a park that borders The Little Belt. It is one of the places where the verse from the national anthem "the mighty beech trees grow along the salty shore" comes true.

[10] "The peace cause" is a Danish branch of the European Peace Movement, founded in 1882 ("*Foreningen til Danmarks Neutralisering*"). The movement agitated against using war as a means to settle conflicts and disputes between states. It was very active in Denmark in the beginning of the 1890s. A petition in 1893 to the Danish government on behalf of the peace cause was signed by 240.000 people (Salmonsens Konversationsleksikon (1919) Vol. VIII, p. 900). See also chapter one: The Danish Context.

appearance of his statements, has a good deal of the fox's nature in him. But perhaps it is with him as with the American editor who had so few subscribers to his newspaper that it could not earn money. He complained about it to a good friend, who advised him that he should write a gross article and be as rude as possible to some other newspaper, that he would naturally receive a response and would have to answer again. This always interests people, and he would get many subscribers, if he could just keep the controversy going. It is possible that the same is true of Wulff, in which case one would have to pity him in his battle to advance in life.

[Also printed in VT, 21st December 1891] I was involved in a quite funny story here a few days ago. One evening, as I sat in my room and worked on the accounts for the plantation, one of the people came in to me and announced that there were two Negro girls who wished to speak with me. I had him bring them in. I asked them what they wanted. They told me that they were slaves [VT edition says concubines] to a rich Arab, who lived 6 mil [30 miles] away on the other side of the river. He treated them so badly, beat them and even put them in chains a couple of times and starved them, because he thought they were unfaithful. They said that he had twelve slaves that he, in his madness, treated the same way. They asked me to buy their freedom or try to convince the Arab to let them work for me for a monthly payment. They asked me to do everything I could for them, because, they said, "our master is a very evil man; and, if he catches us now, he will, for sure, beat us to death; when he gets us back." They looked so frightened that I felt sorry for them, and I said that they could stay here and I would try to talk to the Arab. A couple of days later he came to town. When he learned that they were in the house with me, he went to the fort and made a complaint with the senior officer. I received a letter telling me to deliver the girls to their rightful owner. When the Arab came with the letter, I knew what way the wind blew. I called for my hat and went to talk to the officer, because I believed he had acted without thinking over what he had done. I found him unmoveable. He said that the Arab was a very good man, very courteous and correct. If he did not want to sell his slaves, I had to give them back. I took the liberty of telling him that a good Arab was very rare and that the greatest villain among them would always be polite to the government, since it was necessary for him. I said that I was quite convinced that this Arab was a villain.

When I, however, saw that he would not change his mind, I went back. I sent after a man who lived on the edge of town. I told him that I had two slaves that I, when night came, would send to him and that he was to keep them hidden for a couple of days. He was willing immediately. While the Arab waited in front of the house for his slaves, I had them dress up in my clothes and exercise walking like a man. Thereafter I sent them, one after the other, with one of the house's people right by the nose of the Arab, who did not recognise them. They happily reached the house where they were hidden for a couple of days to the great irritation of the Arab, who set heaven and earth in movement to find them. When he, without success, left, under cover of darkness I put the two girls aboard a sailboat which immediately set sail for Bagamoyo, where they now live happy and well. The Arab cannot understand what became of them. After a month, if you can believe it, he came to me and asked if I would buy them, as he could see that he could not get them back.

It is going very well with the cotton plantation. I believe that the house I will live in will be finished in a couple of months. I am having the trees chopped down, so that I can plant after the rainy season. It is now 4 o'clock Sunday afternoon, and I think I will go to the other side of the river and shoot a couple of wild wood pigeons for my lunch tomorrow. Therefore you will have to be satisfied for now, because I cannot sit still any longer.

With many friendly greetings
Your devoted brother
Chr. Lautherborn
PS: It is unfortunate the invitation to the exhibition came so late. Thank you for the pictures

Deutsch-Ost-Afrikanische Gesellschaft
Kikogwe, 8th October 1891

Dear Parents!
It has occurred to me that it has been quite a long time since I wrote to you. I discovered today, while going through my letters, a letter from mother which I am fairly sure that I did not answer. I have just so much to think about just now until I have the plantation well established, that it can easily happen that, when I get a letter, I read it through and then lay it down some place or other and forget it. The company requires so much written work of me, that they nearly will have an account of everything from Kikogwe. I cannot blame them. They have good reason to be suspicious, because so much money has disappeared here in Africa without any accounting. Because I use about 4000 kr. each month, all eyes are directed at Kikogwe from Berlin as well as Zanzibar. If there is any kind of error in my books, I immediately receive ten to twelve letters from every part of the world asking me to explain the confusion. Since I realise that I am not much of a bookkeeper, I tried with all I had to learn how to satisfy the gentlemen. But then our boss decided to write me a very rough letter reproaching me for an error I had made. I had written a certain amount of money in memorandum instead of in the main book. My patience was now used up, and I sat down immediately and wrote a letter back as coarse as possible. I told him that it was more than enough that I had to be annoyed with the foolishness of

'My house in Kikogwe'

300 to 400 Blacks, but foolishness from a European annoyed me much more. It would, therefore, be a pleasure if he would free me from more letters like the last one. I had other things to worry about than to sit in the office all day and, furthermore, had no contract as a bookkeeper, but as a farmer. He would get an account of every shilling I used, but not according to the double, Italian bookkeeping system, just so he knew.

He could have his bookkeepers rewrite what I sent him. I refused, in the most definite terms, to give any other account than what was customary for any farmer and asked him that if, in the future, he wanted answers to his letters, to write courteously as this was the first and last coarse letter I would answer. That was a month ago, and I have, so far, not heard from him. No doubt he is quite satisfied.

I now live on the plantation in a two-storey house I built myself. On the second story there are three large rooms, five square metres each and a veranda runs along the entire house, three and a half metres deep. Below are the office, bathroom, pantry, kitchen and dining room. The house is fifteen metres long and nine metres wide, and built of coral stone. It lies about one hundred metres from the sea on a little hill, and I can, from my

'My house and my cart'. The man driving the cart is CL's assistant.

window, see all the steamships coming from and going to Europe. There is a wonderful breeze here. The plantation lies on a peninsula, and the wind almost always comes from the sea to me. I also feel stronger every day I am here and have an appetite like a threshing machine. At night it is so cool that I have to cover myself with a wool blanket to keep from being cold. It is similar to Europe. A large building for the storage of cotton, a stall for 40 oxen, a warehouse and living quarters for 40 workers are now ready. There are about 70 td. [app. 90 acres] land under cultivation with the most wonderful cotton I have seen in my life, which I will begin to pick in fourteen days.

All the Europeans enjoy coming here to Kikogwe, and on Sunday I usually have many guests. We enjoy ourselves by walking in the woods and watching the apes spring from branch to branch or by sitting on the veranda and telling hunting stories. Herr von Soden,[11] the Governor of [German] East Africa,

[11] Julius Freiherr von Soden was Governor of GEA 1891—1895. He was the most experienced colonial administrator that Germany had produced, having come from German Cameroon, but was sick and tired. A great bureaucrat he soon tangled with both military and commercial interests, and after one and a half years, he was retired (Koponen 1994:99-100).

No text but it may be the large building for storage of cotton being built and living quarters for 40 workers mentioned in this letter 8th October 1891

'Four of my friends' (unfortunately unnamed)

visited me recently. He never tired of expressing his satisfaction over my work and promised to come again soon, since he had great interest in cotton cultivation. The German warship, *Schwalbe*, here to protect the various places on the coast, came to Pangani a couple of days ago, and the captain, with all of his officers, visited me. They were all very nice and friendly, and I became quickly acquainted with them all. When they left, towards evening, I invited them all to come and eat dinner with me the next day, which they, with pleasure, agreed to. Since a quite ordinary, uncivilised Negro cooks for me, I was very busy the next day giving him the necessary instructions and showing him how he should make the food. I was finished in time, so that I could entertain my guests when they came. Yes, I know that you are smiling hearing about me preparing food, but I have acquired that art here. Just see what my guests had for dinner: meat soup with dumplings, fish salad in mayonnaise sauce, beef a la mode with browned potatoes and vegetables, roast duck stuffed with oysters, and, as dessert, pancakes with jam. With the soup, we drank port, with the fish, Rhine wine, with the duck, red wine and with the dessert, champagne.[12] All the officers were wide-eyed over the dinner and expressed their surprise that I could bring forth such a triumph in Africa. I believe, however, that the pancakes with jam won the most applause, because they ate with an appetite one finds only with sailors and asked me, when they left, to tell them how my cook made the pancakes, since they wanted to make them onboard. I fulfilled their wish, and they promised me that, when I visited them, I would get as many pancakes as I could eat according to my own recipe.

My household staff consists of a cook with an assistant, two Blacks to keep everything upstairs clean and one for below, and a Black girl who carries water in the bathroom and keeps everything clean there. As an assistant I have a botanist (Fischer) from Berlin. Jager, whom I wrote to you about, is no longer here. On the farm there are about 200 Blacks with their wives, who see to it that I have the necessary evening entertainment, as they beat drums and dance nearly every evening. In the beginning I complained about the noise, but I am now used to it and look forward to their cheerfulness. I have recently thought seriously about going to America. I still get letters from Mrs. Gay, and she is nearly crazy to get me back. She says Bates will be glad to lease his farm to me on good conditions, if I would just come. It would, of course, be very good, but, on the other side, I am

[12] This is a menu from Mrs. Nimb's Cookbook. Mrs. Nimb's cookbook is the Danish equal to Escoffier's *Guide Culinaire* and today a classic. Louise Nimb (1842—1903) earned her reputation as a cook at the famous restaurant 'Divan 2' in Tivoli (Copenhagen). She wrote her equally famous cookbook in 1883. According to Karl Larsen's notes from his interview with Caroline Carlsen, CL translated parts of the cookbook into Swahili for his cook to read.

quite sure that DOAG will pay me from 8,000 to 10,000 marks per year,[13] if I will remain when my contract runs out. Therefore I am pretty sure I will decide for the latter, because the pay is pretty good; and, if I lease Bates' farm, it would be less than two months before I was married. Mrs. Gay would see to that. She would then be certain to keep me, and I have, as yet, no desire for that, so I will probably stay in Africa. In any case I will come home first and stay five to six months, and we can talk about the matter. I am, thank God, fresh as a fish and have the best hope for the future.

Live well and greetings from
Your affectionate son Chr Lautherborn
Greetings to everybody!

[13] One German Mark equalled 88.88 øre (Salmonsens Konversationsleksikon, bd XVI, p. 636). 10,000 Mark would have valued 8,880 d.kr. and 8,000 Mark would have valued 7,120 d.kr. in 1891. The yearly salary for a skilled worker in Denmark in 1891 was app. 1.000 d.kr. (Jørgen Pedersen: *Arbejdslønnen i Danmark under skiftende Konjunkturer i Perioden ca. 1850-1913*. København, 1930).

CHAPTER SEVEN

KIKOGWE: THE COTTON PLANTATION

Deutsch-Ostafrikanische Gesellschaft
Kikogwe, 13th November 1891
[Also printed in VT, 23rd, 26th, 27th, 29th January and 11th February 1892]

Dear Brother-in-law and Sister!
Finally I am again able to write you a letter. It must be about five months since the last, but I have had so much to do during this period establishing the plantation, that all desire to write was lost.

As far as I can remember, I last wrote about my arrival here in Pangani, and that I had begun to work here. As I wrote before, I brought many of my workers from Bagamoyo. When, however, according to the company's wishes, the plantation was to be set up on a larger scale than first discussed, it was soon apparent that I had too few workers. The Blacks from Pangani and those in the village of Bweni on the other side of the river would not come and work for me, because they feared they would be treated badly. Work, therefore, went slowly at first. I was happy, however, when three *Nyamwezi* caravans arrived, and I could have three hundred men per day. Among the caravan people, many knew me from Bagamoyo, and it was, therefore, easy to get them to work.

At that time the grass here was about eight fod [feet] high and so thick that one could make way through it only with great difficulty. As it was quite dry, I set fire to it, and it all burned away, but a close network of strong roots and stumps remained. I believe that four strong Danish horses could have pulled a plow through it, but, for the Indian oxen I had, it was impossible. They would have fallen dead over the effort in the severe heat. Therefore, I put 300 *Nyamwezi* to work, each equipped with a hoe, and they chopped all the grass roots away and loosened the earth. The next day the roots, when they were dry, were gathered up and burned. Bushes and trees were chopped down and the roots dug out, and, in a short time, I had a dry, clear piece of land. Now I set six Indian oxen to work with three, heavy, English harrows and harrowed all the land, after the Blacks had chopped away all the grass and bushes. In six weeks I had 50 td. of land [65 acres] as clear and smooth as a field at home and then began to plant cotton, which is planted in rows.

In America one uses a marker to trace out the rows, which is made with an iron axle of 1 3/4 tommer [inches] wide with three wheels. One wheel sits in the middle and one on each side of the axle. The two outer wheels can be adjusted so that the rows will be from one to one and a half metre from each other. Over the middle wheel, screwed to the handle, is a seat. Two horses are harnessed, the marker pole is positioned, and, from the seat one can see the pole that has to be kept in a straight line between both horses. With such a marker one can mark 15 td. of land [19 acres] per day without straining oneself or the horses. Since I do not have any horses here, and, with oxen, such work is difficult, I made a marker from a cultivator by removing all the small ploughing blades except for those in the middle, which are four tommer [inches] wide. Then I bound a strong pole crossover and hammered an eight tommer [inches] nail through it on each end, one and a half metres from the middle. I then harnessed six strong Blacks to it, positioned my pole and intended to plough the first row myself. I explained to the Black in the middle that he should make certain that he had all the poles in a straight line with himself. However, when I had come to the middle of the row and stopped to move a pole, I looked back to see if the row was straight. The very best boa constrictor, in my opinion, could not have made more curves than were in that row. I soon discovered that the Black could not see if he was in a straight line with the poles or a metre to one side or the other side. He looked only at one pole. What happened to the others when he walked crooked did not bother him. Therefore I went in the middle, and one of the Blacks held the marker, so the first row was lined out properly. The next row was lined out with the large nail always in the first and so on, the rows, thereby, all had the same distance. A sixth man followed the marker and laid seed in the groove it had cut. A pair of oxen with a light harrow covered it up. In this manner I was finished with the clearing and sowing of approximately 70 td. [90 acres] of land by the 5th of July. Today, the 13th of November, there are seven fod [feet] high cotton bushes, from which I am fully occupied with picking cotton. From each td. [1.3 acres] I believe I will harvest about 600 pund [pounds] of clean cotton, which one can call good for the first year.

At the same time, I had to see to housing being built for my workers and I had quickly built a long house of strong poles and covered it with palm leaves, in which there was good room for forty workers. I also rented a number of small houses around the plantation, so that all had roofs over their heads. I then began to work on a stone house for myself, which I finished on the 15th of August. I might send you a picture of it. There is a lot of stone here from the old Portugese forts.

I had it all torn down to build with. The work went well, but the caravan people left all at once. Since the heavy work was nearly finished, and I had enough regular workers, it did not hurt me. The people here, however, who are very superstitious, told me that they did not want to remain on the plantation, because it was haunted by the Devil. I knew a young girl had died very quickly. They said that she had been outside the house at night. There the Devil had grabbed her and squeezed her stomach, and that was why she died. I tried to convince them that they were wrong, but I was not

'A house for my machines under construction'

No text and no date. Maybe the long house covered with palm leaves, used for workers quarters, mentioned in this letter 13th November 1891

able to. A couple of weeks later a young woman became very sick with fever at night. In the morning she lost consciousness. The people now told me that they would not stay any longer, for now they were certain that evil spirits were about the place. They said that the young woman would not last long, and, when they had buried her, they would go. This was a serious problem. If the plantation got the reputation among the Blacks that it was bewitched, I could just as well pack my bags, because it would be extremely difficult to get workers. I went to see the young woman and saw that she was very sick. She was quite cold over her entire body. She had a high fever, and, as it often happens here, the fever falls little by little to way under normal. If the sick person gets no help, he or she will quietly fall asleep and die. The first thing I did was to wake her and make her a beer glass filled half with tea and half with cognac. I then gave four women one litre of whiskey, which they were to give her until she became warm again. I had a fire built close to the bed and put a woman at her head to keep her awake and not, for any reason, to allow her to sleep until the sweating began to appear. I then went to the people and explained that the woman became sick not with the Devil's help but like all other people become sick here and everywhere else. They said that in other places they become well again, but here they always die, because the Devil is involved. I told them that the woman was not dead yet and might become well again. "Yes, maybe," they said, "but, if she dies, which is more likely, will you then believe that the Devil is here?" I said no that I did not believe in the Devil. "But, if the young woman gets well, will you give up your belief in the Devil and stay with me?" Yes, they all agreed. "Let's get to work then," I said. "If the woman dies, we can have a meeting tomorrow." They all went to their work, and I went into the house to see the patient. There was still no change, and the women who worked with her all looked depressed. She wanted to sleep all the time, and she had to be shaken hard to keep her awake. At about ten, however, the warmth began to return, and an hour later she was bathed in sweat. I then let her sleep as long as she wanted and, in the evening, gave her a couple of quinine pills. The next day she was well enough to sit outside the house. Belief in the Devil fell significantly, when she returned to work. It had not, however, disappeared completely.

I believe it was in June that we had a lunar eclipse. In the evening, as I sat in Pangani, I heard an ear-deafening noise down on the street. People screamed, beat on tin cans and drums, in short they made a noise that is impossible to describe. I asked one of the people nearby what it meant. He said that the Devil wanted to swallow the moon. "A good piece of it is dark tonight, and, if we do not drive him away with our noise, we will never see the moon again." They kept up their noise until the moon was clear again and went home in the belief that they had driven off the Devil. Often someone or other will believe that there are spirits in his house. He has all who have whistles and drums come to his house. They make music all day, which is,

No text and no date. On the shore coral stone. See description of mortar making in chapter 6

No text and no date. The workers are fabricating bricks

in my opinion, horrible enough to chase off an entire company of evil spirits.

During construction here I had mortar hoisted up to the walls. Underneath there was a man who hung the filled buckets with mortar on the hook and took off the empty bucket. Above him, a man received the mortar and hung the empty buckets on the hook. Through carelessness he let a bucket fall down on his fellow worker, which made a big hole in his head. When I bawled him out for not being careful in his work, he said, "I did not want my comrade to be hurt, but the Devil was mixed up in it. That is why the bucket fell." He could not see that he had not put the hook on correctly, and that that was the cause. A Negro always has an answer ready, when he has done something wrong. Since he is quite lazy, one often catches him loafing. When he is called to account for it, he almost always says, "I was so hungry that I could not work any longer but had to rest awhile." They are always hungry, and that word is probably heard a hundred times a day, even though they earn a fairly good pay here. They receive 10 rupees per month. A rupee has 64 pesas. They can live quite well with 10 pesas per day—a little over 1 rupee per week, but they are always hungry. What they do with their money I cannot understand.

But I have gotten away from the work on the plantation. When the house was finished, I built a stall for 38 oxen, a large warehouse for cotton which makes up one wing, a smithy and carpentry shop and storehouse in one building, quarters for three of my foremen, a shed for tools and storage for cotton balls in one building. The farm is close to the sea, and the cool, sea breeze makes it pleasant for people and animals to live here. I now need to build a building for cleaning the cotton of seed. I hope all will be finished by New Year.

Nine to ten days after the cotton is sowed, it comes up. In the short time of thirteen weeks it has four leaves, which must be constantly cleaned, and the earth has to be loosened around the young plant. I had workers always clean and loosen the earth close to the young plants with their hoes. In the middle of the rows I ploughed with the cultivator, which looks like a three-sided harrow, and kept everything clean there. To get the Negroes to drive with the oxen between the young plants without breaking them was a piece of work I will not soon forget. They all have a natural fear of oxen and are afraid to go up to them and harness the yoke. As soon as the ox looked to the side, they all thought he would gore them, and neither good words nor blows could make them go there. Therefore, in the beginning, I had to harness the oxen myself everyday. Then, while teaching them to use the reins, I explained that when one pulls the rein on the right and loosen the left a little the oxen will go to the right and conversely to go to the left. But it seldom failed that the oxen went in the opposite direction, because the wrong rein was pulled. In various spots they plowed up all the cotton plants, so I had to replant them. But, after I had struggled with the workers for about

No text and no date. In the foreground 'makutis'

a month, they understood fairly well how to drive, and, thereafter, I kept all the cotton clean with the six oxen.

When they realised that they could do their work fairly well, they came to me and asked for an increase in their wages. They said that they now did their work much better than the others. "With the six oxen we clean more land than fifty men with their hoes, and we should, therefore, also earn more. If we did not clean with the oxen, you would have to give a lot of money to others to keep everything clean." I thought, however, that it was the oxen that pulled the cultivator and did most of the work, while they had very little work going behind them. Actually I should pay them less, since they did not have it nearly so hard as the people who worked with the hoes. They did not agree with this, but thought that it was probably better that everything remained as before.

I forgot to tell you that the day after the lunar eclipse I tried to explain the earth's, sun's and moon's position in relation to each other with three, large balls. I did not, however, get very far, because none of them would believe that the earth rotated. If it did all the houses would fall off, and people could not stand on a ball, even if it was very big, when it started to move. They laughed greatly over my joke, as they called it. I saw them talk about it with each other different times during the day. They always laughed heartily over it, as if I had tried to play a joke on them. But it was actually too much to ask that they would believe it; it does sound so incredible.

The cotton grows very fast when it is kept clean—so fast that one can see it. Four months after sowing the cotton bushes here were seven fod [feet] high and closely covered with fruit buds. The rows were one and a half metres from each other, and the bushes are now completely grown together. It looks like they were broad-sowed and not in rows. The Negroes like plucking cotton here as much as in America. The difference being, however, that there they pluck two to three hundred pund [pounds] a day, while here they do not come up to more than forty. I believe, however, that they will be able to pluck one hundred per day next year. The apes are also interested in my cotton field. They pluck the cotton and eat the seeds, then they throw the cotton away.

I go out every morning to chase them away. As soon as they have sight of me, they stand right up on their hind legs to see which way I go. When I come closer, they go over the cotton bushes in great leaps and disappear in the forest, where they again come in view in the high trees. There they spring about making noise, as if they were having a good time over stealing the cotton.

In the last pack of *Vendsyssel Tidende* I saw an article with the title "The Blacks' Revenge," which writes a great deal about the Germans' cruelties here in Africa and about Hr. von Zelewski's last, unfortunate expedition, during which he was killed. I do not know who has written it, but it is clear that the gentleman in question knows very little about conditions down here. I have never seen or heard anything about the cruelties he writes about, and yet I live among the Germans. Hr. von Zelewski's last expedition was not sent out to conquer land, but to punish the *Does*, also called the *Mafiti*.[1] In 1891 they were down here on the coast to rob and plunder the peaceful *Zaramo* people who fled. They took possession of their property. Anyone they caught they killed in the most abominable ways I have seen. All the bodies were mutilated. The women had their breasts cut off and their bellies split open, so that the entrails hung out on both sides. The men were so pierced with spears that they could hardly be held together. The small

[1] Zelewski and his expeditionary force was beaten by Chief Mkwawa of the Hehe people. This victory gave Mkwawa the title 'the Black Napoleon.' CL refers to the Doe and Mafiti as the peoples that Zelewski tried to subdue. The Doe were not involved but the Nguni peoples in the South, including the Hehe, who adopted many Nguni characteristics, were often referred to as the Mafiti during this period. British Naval Intelligence, *A Handbook of German East Africa*, New York: Negro Universities Press, 1920, p. 25. Zelewski was very cavalier about the military prowess of the Hehe and took few precautions about an attack. He and others in his entourage paid for this with their lives (Iliffe 1979: 108-109).

children were stuck through with a spear on both sides of the backbone and hung up between two, forked branches. They built a fire under them and watched with pleasure how the small child writhed in pain until it was dead. The *Zaramo* came and sought protection from the Germans. Three to four expeditions were sent against the *Does*. At first they fought back, but, when several hundred were shot, they took flight. There was no time to pursue them, and, therefore, I believe they will come again this year, in part for revenge for their losses last year and to plunder the *Zaramos*, who, as soon as they heard the enemy was near, fled to the Germans and asked for protection. Two expeditions were sent against them. One was led, as we know, by Hr. von Zelewski, the other by Hr. Lieutenant Smith. Hr. Smith's expedition returned after a couple of weeks had passed without having met the enemy. On the other hand Hr. von Zelewski met and continued to pursue them, which was the reason that he came to, for him, an unknown terrain; where he had to pass through a swamp with close growths of grass on both sides. The *Does* had hidden in the high grass and numbered about 15,000. When Hr. von Zelewski and his soldiers were in the swamp, the 15,000 *Does* [it was in fact the Hehe people who killed Zelewski see note 1] suddenly fell upon them. It happened so quickly that the soldiers did not have time to fire their rifles. They were all stabbed to death with the exception of fifty soldiers, who were somewhat in the rear with the wounded and the provisions along with three or four Europeans. They fortified themselves and beat back the enemy each time they attacked. After a couple of days they began their way towards the coast, which they, fortunately, reached with the sad news.

In another article it is written that the German's methods have alarmed the English. No doubt, the *Times* have delivered that news. But people who live in glass houses should be careful about casting stones. The English have probably forgotten how it went in India, but there are no civilized people who have not carried out such atrocities. In a place like Africa it is very difficult to function with gentleness, but I must say that the Germans are as gentle as possible.

I hope that the crates are now in your possession. I could not pay for them here, and that is why I sent them in that manner.

Merry Christmas and Happy New Year
Yours affectionately
Chr Lautherborn

Deutsch Ostafrikanische Gesellschaft
Kikogwe, Pangani the 8th of April 1892

Dear Parents
I have read Mother's letter again and again, but I still cannot grasp the thought that Regine[2] is dead. It seems like a dream to me, the reality of which, unfortunately, there can be no doubt. I only remember her as a young, healthy girl, who was always ready to have fun with me. I always thought that she would be the oldest of all of us. When I read Father's letter, I felt immediately that somebody was dead in the family, but I never thought about Regine until, to my great sorrow, I saw the first two lines of Mother's letter. For Mother it must be the greatest sorrow and loss, and it saddens me that I cannot be there to console her. I will, however, not travel around to other cities, which was my plan, but take the quickest way home to be with you. I will leave Africa the 10th of May, but I have agreed to travel to America for the company to buy various machines, so I can hardly be home before my 33rd birthday on the 5th of July. I will write to you from every place that I can and try to save as much time as possible. This sad event is only half an hour old for me, so please excuse my short letter. Everything seems to be going around in my head, and it is nearly impossible for me to think clearly. A longer letter will follow, when I have arrived in Rio.

Your affectionate son,
Chr Lautherborn

[CL left GEA in May 1892 and went to America to buy machinery for DOAG, then went home, staying until late October/early November 1892. The following letters/newspaper articles must have been written while travelling or most probably when he was home in North Jutland]

[Printed in VT], 16th September 1892

Life of the People in East Africa

The worker lives in a mud-and-wattle house with a roof made of palm-leaves. His bed is made of palm-fibres woven together and stretched on a wooden frame. He raises from this bed 5:30 in the morning, to commence work at six. Before that he eats cold porridge, made of maize-flour or millet, left over from the evening before. Half past eleven, he returns home. His missus has cooked him a meal. This is her only duty as the bringing up of children does not cost her much effort, their clothing even less. The little Black ones run around as naked as they came into the world. Supper nor-

[2] Regine was CL's older sister by two years, she was only 33 years old when she died (see Chapter 1, The Danish Context).

mally consists of cooked rice with curry, cassava and fish. The afternoon's work lasts from 1:30 to 5:30.

Then comes the evening with its cosy calmness. The Blacks gather in homely cosiness and sit out under the roofs of their huts and chat while the water-pipe chuckles and the cigarettes send their blue smoke out in the quiet of the evening. What occupies them, besides the episodes of the day, are tales about their travels. Many have travelled widely around the 'dark continent,' more of them have even travelled across Africa. It is interesting to listen to what they say in their childish language. Sometimes the chat will also be about the Arabs and the Europeans and their peculiarities. Every new object interests and amazes them. If one of them during the day has seen a pair of binoculars or a burning-glass or another European artefact, it will be discussed with exclamations of astonishment.

It was an object of special sensation when one day they had seen a European take out his false teeth. One of them told, the others listening breathlessly, that one of the whites was a sorcerer who could take out his teeth and hold them in his hand. He added that he would willingly give one month's pay to be able to do the same. If one of the Blacks is a good story-teller he is everybody's friend and people flock around him.

Sometimes they will dance with passion or play around as if they were children, play tag, cops and robbers, and other games we know from the school yard. It is funny to sneak up on them and listen to what occupies them. They have a rare talent for apt comparisons and to catch the character of a man in a nick-name. We had an old officer, with a big and strong chin-beard. He was immediately nick-named 'Mr. Goat'. Hr. von Zelewski, the brave officer who later was killed in an attack, was well known for his strength. His hand hit as if made of iron. He was named 'Mr. Hammer.' A clerk who was rather sedate, was nick-named 'the slow one.' Hr. von Gravenreuth, who crushed Bushiri's uprising was named 'the lion of the coast.' A carefree soul who in his spare-time often drank some bottles of wine, and who entertained himself by throwing the empty champagne-bottles up in the air and shooting at them earned the less respectful 'Mr. Crazy'; a jovial German was simply nick-named 'Mr. Stomach.'

It is even better, though, when the Blacks, as is the custom here, dramatise their critique. Some of them will mock people who do not speak Swahili well. There are two parts. One plays the Negro, the other plays the European, who blabber away in a funny mix of bad Swahili and broken German. The audience rolls over, screaming with laughter.

I saw one of these native artists perform 'Mr. Stomach.' He placed a massive pumpkin on his belly and tied it up with a cloth, where after he waggled around, finally to sit down puffing and blowing, with his eyeballs coming right out of their hollows. Around him people screamed with laughter. One day a very hot-tempered European forgot himself and hit one of the Blacks. In the evening he could see the whole scene re-enacted in the workers' circle as one of them, with agitated gestures, rushed himself at one of the others, and they ended up rolling around in the dirt. It all looked so funny that the hot-tempered man, who in fact is quite good-natured, started to laugh too and from then on strained himself in order not to become part of the evening entertainment again.

The critique may sometimes also aim at one of the men from the Black community. During the uprising we had sailed up the river to punish the rebels. Suddenly bullets whistled around coming from all sides at the same time. We covered up as best we could. One of our soldiers had in his fear ducked at the bottom of the boat, and from here he had shot in each and every direction without caring where he may hit. Some evenings later I saw the Blacks perform the scene where a shaking, huddled up body lay, shaded by two planks leaning on each other, holding a gun which he, without aiming, fired up in the air. One could see the laughter chuckling in the bodies of the others while the sinner sought to excuse himself.

The objects of the native's joking, though, is first and foremost the Europeans. One of the Blacks had been to Zanzibar and I saw him first perform the Frenchman, bowing, gesturing, strutting around talking; soon after he straddled like an English sailor, pretended that he had a large lump of chewing tobacco in one side of his mouth, and cursed and swore. One day I heard one 'Hurrah' after the other. I snuck in on them and saw the re-enactment of a dinner-party the Europeans had held the day before. They sat in two long rows around the table; every moment one or the other stood up, gave a speech, then the others stood up and they toasted, touching their coconut shells with each other. Whereafter, giggling, they would shout their 'Hurrahs' just as strongly as any animated European party could have done.

Apart from that, they want us to know that they know how to behave in a European manner. Whether it is morning, afternoon or evening, as soon as one comes across one of the Blacks, he will salute you in a military manner, putting his hand up to his wooly, curly hair and jovially great you "Guten Morgen" [Good Morning].

[Printed in VT], 20th September 1892

Life of the People in East Africa
(continued)

One of the beautiful features of the Negroes is their love of children. They will sit a whole bunch together and let a baby walk from one pair of arms to the other, each of them showing true signs of joy patting the little one. Another sign that children are treated in a more beautiful manner than in many 'educated' households is that children by the age of eight to ten years may attend a popular assembly and even be allowed to speak about matters in which they have experience themselves, like gardening and the like.

Apart from in the mission-schools nothing much we would call education takes place. The young ones sometimes learn how to write a couple of words in Arabic. Others learn from the Arabs to read a little, but it is of no great significance.

Their natural intelligence on the other side is remarkable. I have listened to people who could speak up to fourteen different African languages; they kept all the words stowed away in their iron-memory. This is so much more remarkable as these languages have often been learned during their travels where dangers and hardships, distress and death were lurking in many forms. Perhaps, though, exactly these circumstances are the favourable teachers. Who is to know?

A strange educational establishment is here and there led by an old woman who teaches the young girls their marital duties. When a young girl leaves this 'school', she will, in solemn silence, take a round and shake hands with people. She is then ready to get married and may be taken home, in a more or less ceremonial manner. Sometimes her intended will just take her on his back and carry her home to his hut, one evening when the dancing is over and the drum silenced. If she turns out not to be well prepared for what awaits her, the husband will seek out the old woman and ruthlessly beat her with a stick.

Apart from that it is amazing with what delicacy man and woman may speak to each other. One evening I watched a pair of lovers from my balcony. He was jealous and violently charging that he could not trust her. She turned her dark eyes towards him and in a soft reproachful voice she said: "It is your own mind that is impure, because I love only you." She spoke with such a convincing force that it touched the listener. Another time, we had had a party. People were leaving when I heard a young man bitterly reproach his sweetheart that she had not rushed forward to greet him although she knew that he had walked the long way from Pangani just for her sake. She answered that her feelings for him were so warm that she would not run the risk of the others laughing at him. But my thoughts rushed forward to greet you she said. And he, he did what every man on earth would have done given such an answer. His grumbling subdued until it turned into a happy laughter.

The natives sometimes are a couple just for a few months. Sometimes though their relationship may last a lifetime. Because a marriage may end soon, the women are reluctant to become mothers and act accordingly. I have had the pleasure on my plantation to see the number of little Black ones grow fast. The relationship between man and woman becomes more stable when work is steady and stable; apart from that I sponsor each child with a gift of 2 kr., so my little colony is growing fast.

The sum may sound insignificant, but pay for a day's work is here around 10 øre, so the gift represent almost three weeks' work, and for the best of my men, two weeks' work. "10 øre as pay for a day's work" you may exclaim and clasp you hands in astonishment, "that will certainly do."[3] Oh yes, it will do and not too badly. In the natural abundance here, food costs only 3 øre a day. They will then have 7 øre in surplus for other necessities, so it is definitely not as bad as it might sound.

When a child is born, the mother as her first act will loosen her own fetish and tie it around the neck of the child, until she has had one made for the child itself. In order to get that done she will visit a man who can write a letter for her. This letter is then sown into a little leather purse and hung around the neck of the child. Then the mother will stop worrying about the fate of her newborn.

Sometimes superstition may take on a more threatening shape. One day I heard loud screaming from the plantation and saw one of my men dragging the medicine man along. The medicine man, now looking rather pitiable, had threatened the men with transforming himself into a lion and eating them if he was not given substantial gifts. The commanding officer had had him caned, but the Blacks were of the opinion that it was in their best interest to disable him, just in case.

Now we were in a mess. I therefore took on the air of an expert and demanded to see what he had in the pots hanging in a belt around his waist. He handed the whole arsenate of trash and incense to me. It was just red clay, mashed insects etc.. I examined it solemnly, and after my examination I declared in a firm voice that with these items nobody would be able to transform himself into a lion, only into a pig. After that I took the man aside and gave him the good piece of advice to disappear as fast as he could to prevent any further unpleasantness. There was no need to tell him twice. Never in my life have I seen a man run as fast as this sorcerer. I am sure he set a record any sportsman would envy.

[Printed in VT], 21st September 1892

Life of the People in East Africa
(continued)

The blacksmiths come closest to forming a trade. In peacetime they make hinges and hoes to use in local agriculture. In wartime they fabricate spears and arrows. Their tools are very simple. They sit on their heels before the anvil, a four-sided piece of iron fixed in a wooden block. The bellows consist of two skin-sacks, alternately drawn out and pushed together. We have also people who mainly work as masons or carpenters, but tailors, shoe-makers and bakers are only to be found in more advanced cultures.

The caravan people form a separate group of workers. A caravan nor-

[3] Pay for a day's work for unskilled men was in 1892 in Denmark around 225–250 øre.

mally consists of several hundred people, mostly carriers. Even people of distinction have to carry their own kitchen utensils etc. in a caravan. Donkeys are used as pack animals, but they cannot carry that much or else they risk getting stuck on the narrow pathways in the forest or detained in the high grass. Even people can have a hard time pushing through. Sometimes they may have to walk on all fours to get through the jungle. At other times the caravan people will march through the high sea of grass, no wind at all to cool them, panting, sweating and gasping for water. In this manner they will have to march four to five English miles a day and when they finally reach their camping ground they have to go hunting for meat and search the forest for fruit and nuts in order to have something to eat. If they find a camping ground with abundant food, they will stay for six to eight days. But if the environment is poor or the scoundrel slave-traders have ravaged the area, they will march on as fast as possible as hunger may threaten on this stretch. They will have to make do with the dried strips of meat, millet etc. they have brought along.

These caravans are in fact co-operative undertakings. In the interior elephant-tusks are bought from hunters that live further up in the country. The caravan-men will bring the elephant-tusks out to the coast.

A normal caravan of several hundred men will normally carry about the same number of guns, besides every man has a spear and often a club. Especially on the march back from the coast, where they have traded their precious African objects for cloth and metal wire, the caravans are often attacked by the native tribes who demand customs be paid. In these areas customs have maintained their original feature as organised robbery. A couple of women will follow the caravan as well, as the more wealthy men will bring a wife or two along.

A march to the interior and back normally takes four to five months. When the caravan reaches the coast, the appearance of the caravan-men bears witness to the hardships of their travel. One can count every rib-bone and see the two bones in their arms and legs. Once on the coast, the porters try to find work with the Indian merchants or in the plantations, whereas the men of distinction will do the trading. It will not be long before the caravan folk are glistening fat, like stuffed vultures. They are definitely not squeamish. I have seen them arrive when animals are slaughtered. They will eagerly grab the stomach of an ox, even if it has fallen on the ground, dust the dirt away, and eat it raw as if it was a great delicacy. When they walk in the fields they carry a long, slender stick, in their hair or their fez. If they, while they work, see a grasshopper they will quickly spike it and then continue whatever they are at. Before noon they may have caught a whole meal. I must admit that my mouth does not water when I see this collection of wriggling, green little bodies, but the food is more solid than one should think. The African grasshopper's behind is as thick as a little finger. The Black children will often play with the grasshoppers, pretending they are cows and tethering them up with a string.

After the ivory has reached the coast and has been traded for cloth, etc., it is time to leave. They will leave their low, bee-hive like huts. The caravan organises itself, and starts to move towards the interior. When the caravan moves through the high grass it looks like a giant serpent. The usual perils await. Among these I have not mentioned the crocodiles, which greatly hamper crossing of the rivers. I once saw a caravan cross a river. The men were ferried across, but it was not possible in this manner to spare the cattle, which they had brought in great numbers, from the crocodiles' attacks. Although good riflemen were posted on the shores, these greedy beasts grabbed at least twenty-five cattle, which suddenly disappeared in the murky water with a moan still resounding in my ears. One of the crocodiles, however, had a strange accident. In spite of the horns, it insisted on grabbing the oxen by its head. All it obtained was to get its teeth entangled in the rope. A strange fight developed between the crocodile and the oxen. The crocodile wanted to drag the oxen down under water, but the oxen had gained a foothold in the riverbed, and tossed its head into the air to get free. This way the crocodile was almost towed onto the bank, where the bullets started dancing around it and it disappeared as soon as the rope broke during their fight.

Through perils and dangers like this the caravan will finally return home after months of strained marches, to see their family again and to announce all over the area that they have goods to trade, where after people come from every corner. The elephant hunters will turn up too and once more they will trade.

The elephant hunters kill their prey in a peculiar manner. Before they go hunting they equip themselves with clusters of bananas. They carry these on their head. They also equip themselves with bow and arrows and a little calabash with red ant poison or juice from poisonous plants. Then they sneak into the forest, and try to get close enough to some of the huge animals to be able to shoot three to four poisonous arrows into their legs. Although the elephant may barely notice these pricks from a pin, death is sure. The hunter patiently follows the elephant for some days until it lies down when it starts to feel sick. The hunter will wait until it has fought its last strife, then the Black man comes out from his ambush, and chops off the teeth, which are often so big that it takes two men to carry them. The hippopotamus is killed with a poisoned spear, as long as a fishing spear. Other kinds of game, e.g. wild beasts killed for the sake of their skin, have to be killed with blank weapons, because the hair would fall off if poison were used.

The conditions of the slaves I will tell about another time. The slave-hunters are the curse of Africa, and it is with a bleeding heart that one has to say no to a human being who begs to be bought free. Many will pay the ransom, and the man will later work to pay it back. The ransom is the usual market price, 150 kr. for a man, 120 to 300 kr. for a woman, according to age, beauty and strength. One has to be on guard making such a trade as the Negroes sometimes will try to cheat on you. Some time ago we had two brothers who for some time kept a profitable business running; the one would sell the other, who, as soon as possible, would steal away whereafter the two brothers would share the ransom.

Kikogwe, 25th November 1892
[Also Printed in VT] 5th and 7th January 1893

Dear Brother-in-law and Sister,
It is really about time that you heard something from me. I have also often wanted to write, but have not had the time. The trip here was very pleasant, when I exclude the last days, when the ship rocked so much that one nearly had one's ribs knocked out on the edge of the cabin's bunk. In Naples I met a couple of old Africa hands, and we walked about while we waited for the steamship. We had heard on the way down that eight men had died of cholera onboard, after the ship had left Hamburg. We were, of course, agreed not to travel with the ship, if this were the case. But it turned out a couple of days later that the rumour was made-up in Lisbon by a Portuguese steam ship company, which had a ship going to Africa at the same time as the German ship. They were trying, through this lie, to get as many passengers as possible, since, naturally, no one would sail on a cholera infected ship. We were all very glad, when we, at last, saw *Reichstag* sail into the harbour and shook hands with our old acquaintance Captain C. His healthy appearance was enough assurance for us that everything was as it should be onboard. We lifted anchor at six in the evening and softly slipped out of the bay, which was full of small yachts. We waved farewell to the brown-eyed beauties, many responded, but hardly with the same feelings as ours. We were, after all, saying farewell for three years.

The next morning I had a fever. I do not know the reason, but I had to take medicine and cured myself to the great irritation of our doctor. He believes the medicine I took did more harm than good and that I should always consult a doctor, when one was near. When I, however, was well enough the next day to show myself in the dining room, I believe he changed his opinion. During the cruise in the Mediterranean one swallow after the other came onboard. They were so tired and exhausted that they fell down from the mast, where they had sat to rest. We caught cockroaches and fed them to the birds and gave them water to drink. After a couple of days they were so tame that they came and ate out of our hands.

We reached Port Said after four days early in the morning. The pilot who came out to us would not come onboard, but tied his boat to the side of the ship and, from there, gave his steering instructions. When we had come into the canal outside the town, we were not allowed to come ashore. They were afraid we had cholera, but when the doctor came onboard and saw that we were all well, we were permitted to leave the ship. Port Said has expanded quite a bit since I wrote my first letter about it. Transport through the canal has increased, bringing in colossal amounts of money, and, so that it can get even better, the canal is to be expanded so that two ships can sail side by side. At this time it is so small that one is amazed that a colossus like our ship can go through without getting stuck. I do not believe that it is wider than Østergade in Hjørring.[4] If it went through a beautiful landscape, it would be interesting. But, as far as the eye can see, one sees only sand. The various stations built close to the bank appear, therefore, beautiful with the green trees and bushes, which are planted around them and watered every day to keep them alive. Many families live on the water. An old barge is used as a foundation, and a house is built on that. When they are finished with one place, they tie up to a steam ship and go to another, where they drop anchor as long as the work lasts. Most of them are the people responsible for dredging and repairing the canal. After twenty-four hours of cruising we were out in the Red Sea and had to pay, I believe, 12,000 Fr. to come through. Outside of Suez there are a huge number of sharks. One can see them around the ship, when it is still.

Day by day, the cruise became warmer and warmer. We slept up on the deck at night. On the coast of Africa I saw dust devils rise up to a monstrous height and fly by us one after the other, as if playing tag. As soon as they came to a slope, where an air space was created under them, they collapsed.

It looked like they suddenly disappeared. It has long been a mystery to me why the Red Sea is called 'Red.' Its water has always looked as blue to me as all other water, but the mystery was cleared up on this cruise. We sailed an entire day through a material that resembled red clay. The wind carries the fine sand, partly from the coast of Africa and partly from the coast of Arabia, out over the water, where it falls down and remains on the surface. When one sees it from a distance, it looks like land that has suddenly rissen from the sea. After five days we reached Aden two hours before sundown. It was quite late to go ashore, and I did not have any great desire to do so. But the passengers who had not been there before asked me to go with them. We took a carriage to Aden—the place where one first comes ashore, where all Europeans board, is called Steamer Point. From there it is a good Danish mil [4.6 miles] to Aden. It was a Jewish holiday. They had their long curls around their ears beautifully dressed and, in clean clothes, a novelty for these people, they rode out to Aden. A large square in the centre of town was illuminated and filled with tables and chairs, where they took refreshments. For us the holiday was an annoyance because we want-

[4] Østergade in Hjørring is app. 20 meters wide (see also endnotes for chapter one).

ed to buy ostrich feathers. We tried to get a couple of them to sell us some, showing them a couple of English gold coins. It was too much for their greedy money souls. They asked us to wait for them at a certain place, where they would bring their goods; so that their friends would not see anything. We went there, but there must have been too many difficulties, because they only came when we were ready to drive back and could wait no longer. It is horribly hot and dry in Aden but very healthy. Europeans look as well there as at home, and I do not believe that there is any fever.

After we had left Aden and had come to the Indian Ocean, it became cooler and we enjoyed ourselves up on deck in the evening by dancing. One of the stewards played for us, and I heard the old, known Danish melody from Hjørring, to which the Germans sang:

> Male, Male! Is mein schöne Male da
> Male, Male reisen nach Ost-Afrika.
> [Male, Male (most probably a nickname, maybe for a girl named Malene) is my beautiful Male there, Male, Male we will go to East Africa]

When we were tired of dancing, we went into the dining room, which had a very good piano. In Naples we had bought quite a lot of Italian songs, which we practiced, and which, in the end, went quite well for us.

Finally came the day when we crossed the line (of the equator). The captain was not really well, and the baptism could not be done properly. It was the third time I had passed the line and did not expect to be baptized yet again. Early in the morning, however, I heard the other passengers pouring buckets of water on each other. Since I imagined I would not be spared when I came up, I put on some dirty clothes that I was going to have washed and went out on the deck. All the passengers were there, each equipped with a bucket of water. I had hardly left the last step of the ladder before it rained down so much on me that I could hardly breathe. The sailors were washing the deck, and the boatswain held the water hose, which had pressure from the steam motor, in order to give the water enough force to push all the dirt. I gave him a wink, and he gave me the hose. I sent a jet in between the passengers, which made them shout. They tried to get away, but, since the deck was wet, three of them fell their full length. The doctor was one of them. He had his blue uniform on and was still dry, but, after I sent a jet up his back, there was not a dry spot on him. The other two who fell received the same treatment, and later I had the pleasure of paying the others back. At lunch I brought my complaint to the captain about the boldness of the young landlubbers in baptizing an old and experienced sea dog like me. He punished them by making them give a barrel of Mündener beer in the evening, which we drank to good companions.

We were now quickly approaching our goal. One and a half days after we had passed the line, we came to Tanga. There I met several old acquaintances both White and Black, whom I was glad to see again. I ate dinner in the government building with the former director of our company. The next day we sailed for Zanzibar, where I stayed for a couple of days and sailed on a *dhow* to Pangani.

At the plantation I was received with jubilation by my Negroes, and I, naturally, had to go out with something good for them to eat—rice and a couple of goats. The Europeans looked healthier than they usually do in Africa. The reason for that was that the well, which I had begun before I travelled home, and which was now finished, had ferruginous water. This is very healthy, since blood easily gets thin out here. I have had a horrible amount of things to do since I came here. I have set my cotton machine into operation and have cleaned 5000 pund [pounds] of cotton. It is, however, only a tenth of what I have, and hope, therefore, to be able to please the company with a large shipment here from Africa, which will probably be the first.[5]

The entire village of Bweni paid me a visit the other day with both of the chiefs with all of their wives and slaves, about 300 men in all I believe. They had brought music and performed a dance with singing in the middle of the garden.

"Bwana Malete is here again from Europe. He cleans his cotton with steam now. He himself is as strong as a steam engine, because he has eaten his mother's good food. She had not seen him for many years, and therefore, gave him the best she had. But it is good that he is strong, because he uses his strength to work and not to beat the Black people. We hope that he may be long among us."

When the dance was finished, I passed out a couple of cases of mineral water among them, which they gladly drank. To the chiefs I gave the clothes I had brought, which pleased them greatly.

It is about 28 degrees Rm. these days, but I must say that it is more pleasant than the sharp winds that blow through Hjørring's streets or also down near Lønstrup. I often think it is strange that I tolerate it so well. I feared that the first six months in Africa would be boring after my six months holiday in Europe, but I have so much to do that I have to get up at four in the morning and work until six in the evening to be finished. It drove away all sad thoughts, and I am now quite well. Enclosed I have sent my photograph in tropical clothes as promised.

Friendly greetings to all
Chr Lautherborn

I have written to Mrs. Gay about Hjalmar and will send you her answer when I get it. Tell me if the things I sent Astrid have arrived.

---

[5] In 1893 the Kikogwe plantation claimed to have planted about 276 acres, of which 134 were cotton (Sunseri, 2002:13). CL also encouraged small farmers to produce cotton and to process it at Kikogwe.

Kikogwe, 8th January 1893

Dear Sister!
I have written to Cathinka[6] and Thorvald and now I come to you.
It was an overwhelmingly great pleasure for me to receive the almanacs and the photographs, and along with it was an entire pile of *Vendsyssel Tidender,* so you can believe that I will hold Christmas. Father wrote me a long letter with greetings from everyone in Aalborg, and, a couple of days ago, I received a letter. Can you guess who from? From Miss Chrestine Jørgensen. She writes a very good hand, sent me two pictures from Aalborg, very thoughtful of her. She also tells me that she longs for me and that I must come home soon without first falling in love with a Black. I think she is a very upright young girl. Your letter, along with the almanacs, put me into a very good Christmas spirit, as good as it can be when it is 28 degrees Rm. But, even though I would like to be with you at Christmas, I do not think I would like to be in such cold weather. It was cold enough for me when I was home, and I don't think there was one day in Denmark where I felt as good as here today even though I sit in shirt sleeves and the sweat rolls off of me. Did you notice how the stars sparkled in the Great Bear Christmas Eve? After I ate, I sat outside the house, smoked my cigar and cast a glance now and again at the Great Bear and sent you a greeting.

Christmas is not all that big here. We had some really good food and drink. I had baked some Christmas cakes and biscuits, which were well received. Maybe, when I have time and opportunity to practice, they will become better.

I have also been very busy since returning. I have now cleaned all of my cotton on the machine and pressed it and will, at the first opportunity, send it to Germany. I have 20,000 thousand pund [pounds] of clean cotton, and I believe that the company will be happy for it. We have had, during this so-called dry season, a couple of very strong showers, and everything is as lush and beautiful as can be imagined. The large mango trees look especially beautiful, as they now hang full of fruit. I was on a wild boar hunt the other day, shot and wounded two and had to retreat up a tree, because my case of hardened cartridges had fallen through a hole in my bag. I had a few cartridges with duck shot, and I bombarded the boars from up in the tree until they ran away, and I could go home. One boar did not, however, get far. It collapsed after about one hundred steps. I had my people bring it home and had the reward of my expedition, because the meat tastes quite wonderful without a trace of fat. I am now building a large hen house out in the field, so that the hens will have enough to eat without it being nec-

[6] Cathinka was CL's youngest sister, two years CL's junior.

'At Kikogwe. Herr Nietzch, Frau Hollmann and I'

essary to feed them. Close to the house they completely ruin the garden. Poultry here is the most important source of meat for Europeans. One cannot salt here as at home, and a goat or a cow is too big to eat in one day.

In Pangani there is always meat, but it doesn't hang in large pieces as with us, but is brought to market in sizes comparable to a fist, perhaps a little larger, to a weight of four pounds, and it is impossible to say what one is buying. I know that, on several occasions, camel and hippopotamus has been sold as beef. Both animals are, of course, plant-eating, and I, for my part, have nothing against eating meat from them. But one does not know how such an animal died, certainly not from a knife. It could be dead of some sickness or other. The Blacks do not like such things, but they will sell them. I believe a number of Europeans have met their death here because they did not pay sufficient attention to what they ate. A Black cook is given a couple of silver coins to go shopping, and no one worries about what he has bought before the food is on the table. To inspect the meat when he brings it home does not occur to anyone, and most people are too lazy to build a place where they can have their hens, ducks and doves. They would rather buy an old, dry rooster from a Negro and have the pleasure of chew-

No text. Maybe the European colony in the Pangani area on a visit to Kikogwe, mentioned in letter 8th October 1891

ing on it. Afterwards they complain that the food is bad here, or they eat the expensive and poorly sealed canned food. When someone comes here [to Kikogwe] they always wonder over all the fresh food we have to eat. When we are finished at the table, I take them out and show them my garden, my hen house and cows. They then say that they would like to do the same, but it goes no further than that. *Aller Anfang ist schwer* [All beginnings are difficult] as everyone knows.

The best is, however, that the entire colony drinks only beer or whiskey and soda, because they say the water is not fit to drink. Since the well was finished, we drink only water all day and have no problem with it to the great wonder of all. I keep the well very clean and make sure that only one man draws water, and all the others fetch it. I have also covered it, so that it is impossible for filth to get in. Most Europeans live with the open, Arabic wells, all of which are filled with old branches and, now and again, a couple of dead frogs and the like. Rather than put such a well in order, they drink beer or other things and use water for washing or cooking.

Well, that is enough about food and drink. You are probably tired of hearing about it. One of my workers, by the way, is getting married today

CL stands in the front row, number three from the right with what looks like a leash for the dog in his hand

and they whistle and beat on drums out there and dance until it is light. I have been, a couple of times, encouraged to go out and may do it.

Happy New Year
yours affectionately
Chr Lautherborn

Kikogwe, 14th May 1893
[Also printed in *VT*, 6th and 8th July 1893]

Dear Brother-in-law and Sister!
Yes, it is now a long time since I have written, but I have truly had my hands full, so there has not been too much time left over. As you know all the stock had died when I returned, partly because of poor treatment and partly because they did not have good water to drink. When the last problem was solved, I acquired fourteen oxen and three cows with calves, which, so far, are doing well and are big and fat. The three cows give so much milk, that I had to make a new churn. The old one was too little. To make

the churn I took a glazed drain pipe, which was wide in one end and narrow in the other. I cut it in half, set a cement bottom in, made a lid of hardwood, drilled a hole in the middle and made a stamp. It is the same kind of churn we had out in the country in the old days. You move the stamp up and down, when you want to churn. The Blacks laughed at what was for them a new invention, but now like to make butter in this way much more than shaking it in a bottle as before. My butter is very famous here now. Every time I have visitors, they want some bread and butter. It is something one never gets here. When butter is put on the table, I can be certain that there will not be much left.

But I was to write about the Sultan of Zanzibar this time. The picture I sent you is of Sultan Seyyid Ali bin Said,[7] who is now dead and gone. I should begin with the first sultan, who was here when I came to Africa. According to what people say, for there are no histories about it, the East Coast of Africa was ruled by the Portugese three hundred years ago. How long they had their colonies here I do not know and cannot, with my best efforts, find anything definite about it. But it is known that a powerful Arabic clan from Muscat, to which the present sultan also belongs, carried out a campaign against Zanzibar, forced out the Portugese and, later, conquered the east coast down to Kilwa. Today you can still see strong Portugese forts with round bastions. Even here on the plantation there were many ruins originating from that time. I used the stones from them to build new buildings. The first Sultan in my time was Seyyid Barghash bin Said.[8] He caused the war here. He signed a contract with our company and gave up customs duties on the coast and his possessions for four million rupees. The Arabs became so bitter over this that he, for his own safety, had to side with them and do all he could to make our stay on the coast impossible. I have described the war a couple of years ago. The throne here does not, as with us, go from father to son, but from brother to brother. Only if there is no living brother or cousin when the sultan dies does it go to his son.

By the way, while I remember, I do not know if I have written this before; the first time I travelled down here there was, among the passengers, a princess from Zanzibar with her daughter. She was a sister of Seyyid Barghash bin Said. In her younger days she had fallen in love with a merchant from Hamburg, who had a business in Zanzibar. She ran away with him, first onboard a ship of the English navy for protection against the Arabs who pursued them, and later they took a passenger steamer to

---

[7] Sultan Seyyid Ali bin Said (1791-1856) was the founder of the Busaidi dynasty of Zanzibar. Under his leadership the island became the center of Omani commercial and political activity.

[8] Seyyid Barghash bin Said (1837-1888) was the third Busaidi ruler of Zanzibar. Concentrated on resisting encroachments on his domain from Oman and expanding Zanzibar's commercial activity on the East African mainland.

Hamburg. There they were married, after she had been baptized into the Christian religion. They had three children, two sons, who are officers in the German army, and one daughter. She was going to Zanzibar, because she was going to try to get her inheritance, because, after her husband's death, she did not have much money. The old Kaiser Wilhelm had helped her to raise her children, because she was of royal birth. She did not, however, have much luck in her journey. Her brother, Seyyid Barghash bin Said gave her no money, because she had, by running away, marrying an infidel and taking his religion, so seriously broken the Koran's law that he could not help her without sinning against Mohammed. That he would not do, so she had to leave without success.[9]

Seyyid Barghash bin Said was a very wise sultan, probably the wisest Zanzibar has had. He had, built a large aquaduct during his reign. From a hill about fourteen miles outside of town he had the water run into Zanzibar through pipes. Where before they had bad smelling water, it was now fresh and clear. He introduced electric lights, ice machines and much more.[10] He was highly thought of by the Blacks. The Arabs, however, did not like him. He was too wise for them and stopped too many of their schemes. But it is not correct that he was poisoned. Some say that his brother, who also wanted to be sultan, and others said the rich Arabs, had paid his cook to mix poison in his food. I had just arrived in Zanzibar when he died. As soon as the rumour of sultan's death spread, all of his subjects began to howl up to the skies. Especially the women of the harem howled to such a degree that they overwhelmed all others. It is the custom among the Arabs, and also the Blacks now, that, when a man dies, all of his friends and acquaintances gather in his house and cry for seven days. Actually one cannot really call it crying. It is more a kind of howling, like when cats go courting. The women especially do it. One seldom sees men take part in it. In Zanzibar, from most of the houses, one also heard, until the seven days had past, a kind of moaning, as if someone suffered from severe stomach pains. The sultan was buried in the garden close to his palace, where his forefathers lay.

I say palace, but that is not really the right word for it according to our concept. It is a large, whitewashed building, four stories high, with a veranda for each floor resting on iron posts. Around the verandas are written texts from the Koran with Arabic letters. The roof is of corrugated tin, and, up on the top, there is a small lookout post, so the sultan can be the first

---

[9] This is the famous princess Salme, otherwise known as Emily Ruete (1844–1924). She published her own memoirs in Germany in 1886, *Memoirs of an Arabian Princess*. They were soon published in French and English as well. See also, E. Van Donzel (ed.) *An Arabian Princess between two Worlds,* Leiden: E.J.Brill, 1993.

[10] The ice factory was in existence by mid-century (Eric Gilbert, *Dhows and the Colonial Economy of Zanzibar*, Columbus: Ohio, Ohio University Press 2004).

to see the new moon when the *Beiram* festival begins (Arab New Year). The rooms in the palace are large and spacious, but not equipped with grandeur or taste. A large chandelier hangs in the entrance. On the walls there are rows of polished rifle mounts, where the guards' weapons hang. On one side of the lowest veranda there is various equipment covered with a canvas, on the other, old cannon. On the ground by the side of the building, there are old steam engines, Boyer railway tracks and iron posts left over from the construction of the building. The main street, which goes right by, is not wider than Doktorens Gyde[11] in Aalborg. The customs house is directly in front, and on the street there lie hundreds of sacks with cloves, copra, copal and rubber. The first two send out a strong, unpleasant aroma. Up in the rooms there is European furniture such as sofas, commodes, chairs, and consoles. Long mirrors and clocks hang on the wall, but one gets the impression that they are placed there to fill up space. The audience hall is large and spacious, covered with Persian carpets. The walls are closely covered with long mirrors and clocks. On both sides stand a long row of chairs and above a kind of throne.

The sultan sits here when Europeans, on festive occasions, come to wish him well. Then one is served black coffee and dates followed by a kind of sugar water, and, in conclusion, you get a couple of drops of rose oil on your handkerchief, which smells for an entire month afterwards—even after several washings. When you rise to go, the sultan goes to the entrance and gives everyone his hand in farewell.

When Seyyid Bargash died, his brother, Seyyid Kalifa,[12] came to the throne. That was precisely when it was not known if Zanzibar would go to the Germans or the English. Seyyid Kalifa could not, however, like the English. On the other hand he had a very good opinion of the Germans and would have liked to put himself under their protection. He, however, did not get that far, because he died suddenly, and later, of course, there was a treaty between Germany and England. The Germans got Helgoland and should not, therefore, make any demands for Zanzibar. Seyyid Kalifa's death was very suspicious. He was a healthy and strong man and only sick for a few days. Everyone said that he was poisoned by the English. They did not like his German leanings, and, they had to act quickly; they paid his doctor to get rid of him. There is naturally no clear evidence, but that would also be impossible to obtain. One knows, however, that the English stop at nothing to acquire new colonies. Why not also kill a poor sultan, when it could help

---

[11] Doktorens Gyde is an old, narrow lane in Aalborg app. 3,5 metres wide. It was in the 1880s a street where paupers and day-labourers lived, so it is definitely not to flatter that CL offers Doktorens Gyde as a comparison.

[12] Seyyid Khalifa (1854-1890) was the fourth Busaidi ruler of Zanzibar who relinquished control of the mainland to the Germans and the British.

them. One thing is certain, all of the better Blacks and Arabs I know say that the English were guilty of his death.

So came Ali bin Said on the throne.[13] He was a great friend of the English. The English gave him an honour guard, which had secret orders to shoot him down if he caused trouble. In the beginning Seyyid Ali was very satisfied. He soon noted, however, that he was only sultan for the sake of appearances and saw his error in not placing himself under German protection. He apparently also expressed his feelings in words, which may have cost him his life, because he died as suddenly as his brother. It is true that the English claim he was sick for a couple of years. But I do not believe this, because I had seen him a couple of times and he looked strong and healthy. In Zanzibar the rumour about his death was not believed. I am sending you a piece from the paper, so that you can read about it yourself. Seyyid Hamed bin Thwain is now sultan in Zanzibar. He is the fourth in the five and a half years I have been here. A great deal is written about it being dangerous being the Czar of Russia, but it seems to me that it is worse being the Sultan of Zanzibar. Now, in addition to the present sultan, there is a cousin of Seyyid Ali, Seyyid Hamoud, and a son of Seyyid Bargash, Seyyid bin Khalid left. But I am convinced that, before I turn my nose home again, there will only be one left, and the English will be rulers on Zanzibar.

The sultan's daily work is to hold court every morning. In the afternoon he prefers to meet one or another of the leading Arabs on his estates about the most important work that has to be done. Around about the afternoon he likes to drive out. His carriage is drawn by six horses driven from the buckboard. A company of Bedouin horsemen ride in front. It all looks like how one could imagine a gypsy chief with his subjects. Among the horsemen some ride at an amble, others at a trot, and others at a gallop. The carriage horses are ridden at a brisk trot, and, because the driver holds the reins like a farmer back home, one or the other is constantly springing over to a gallop, which looks horrible. All in all one cannot help but smile over such a parade, which is not one thing or the other. It can immediately be seen that no one is interested or understands how to drive or ride. One thing is certain; he has good material. The horses are beautiful. Now and again one sees Arabians among them. But none of it helps, when they are not handled correctly. The evening, the night and most of the morning, after he has been to prayers, he spends with his wives. I have, a couple of times, seen him in the great hall, which faces the harbour, early in the morning. If an Arab comes by, he looks down at the ground and does not dare look up, because the ladies all have their faces uncovered. With Europeans, of

[13] Ali bin Said (1855-1893) was the fifth Busaidi ruler. He was little more than a puppet torn between the British and his Arab advisors.

course, it is a different matter. We would much rather see a beautiful face without a veil than with one, and one should not let such an opportunity go by. Naturally I did not look down and had the pleasure of many a beautiful face. The owner of which, however, quickly withdrew and locked the shutters when she spotted me, if the sultan was in the hall. Otherwise they were not so careful.

With this post I am sending you a couple of Indian and Chinese trays and frames. Let the man first make a model of them, and you can then have them. I will, when possible, send more. I have received a letter from Mrs. Gay saying that she is willing to receive Hjalmar when he comes. I am attaching the letter. Since you can read English, you might enjoy reading it.

Best regards to all
Your affectionate brother
Chr. Lautherborn
PS: Tell Astrid that she should try to improve her writing.

[Letter to Vilhelm Carlsen, 30th September 1893, regarding possible trip/work in Texas for CL's nephew, Hjalmar, has been left out. The letter contains practical information about ticket prices, climate and therefore what clothes to bring, etc.]

Kikogwe bei Pangani
Deutsch Ostafrika, 21st January 1894
[Also printed in *VT*, 29th and 31st March and 2nd, 3rd and 7th April 1894]

Dear Sister
Your letter of 2nd December 1893 reached me here between Christmas and New Year. It was a pleasure to read, because the tone was well suited for Christmas. I can very well understand that you, in the midst of your flock of children, can have such thoughts, but you describe it, (child-raising in love) with an enthusiasm as if the thought was yours originally, and it is not actually. At the teachers' meeting in Aalborg, when the priest with the black hair spoke so beautifully about the same subject, I believe we both agreed that we completely shared his point of view.

However, it is my opinion that raising our own children in love is easy for us, because we love them as our own flesh and blood. But to love our fellow men and seek to raise them in the same way is immeasurably difficult. How often, in life, do we not meet people with a character that fills us with horror and disgust? One, of course, tries to handle them as well as possible and overlook their failings, but to love them is, I think, asking too much. One forces oneself to be good to them, but I do not think one feels love. You write about the labour question and believe that it is unjust that the small man is oppressed by the big, who lives at the expense of the other. But is it not that way in all of nature? See, for example, the shark that swallows all the smaller fish, because it is the stronger, the eagle the dove, the wolf the lamb, the lion the deer, and go into the jungle and look at the enormous trees with their large crowns taking all the sun and light and, under them, small trees that try to grow up But, when they reach a certain height, they wither and die, because the big ones take all light and nourishment from them. It is the same with us. We are, however, equipped with the ability to think, which nature has not bestowed on the others. Therefore we feel it is unjust that one should live in luxury, while the other does not have enough to satisfy his hunger. We try, as best we can, to relieve the need of the poor. I do not believe, however, that you are right in your opinion that the capitalists cannot, in complacent satisfaction, enjoy the goods bestowed on them in life, because the workers' complaints and grief lay a curse over them. As feeling human beings, when one sees misery on one side and abundance on the other, we have a tendency to judge the last too hard. It is not known how many capitalists have come into their money in a completely justified manner, or how many workers, who when the week is over, go to bars and drink up nearly all they have earned, and the wife and children at home must suffer hunger and need. A competent, sober worker does not suffer need in life, unless some serious disease or other holds him down for an extended time. If he is known as a good man, there are always people who will help him, until he gets back on his feet. You must not think that I always approve of what the capitalists do. I know very well that they often, with their speculations, drive prices on things that the workers use sky high, while they cut wages. It is just that I think that both sides bear guilt for the one's misery and not just the capitalists alone, as you stated in your letter.

We had a very enjoyable Christmas here. We had a beautiful Christmas tree in Pangani, with many, wonderful gifts. However some were funny more than useful. I received, for example, a box with nails, gimlet and file, which I am supposed to use to build up Africa. A doctor received a large box with a mass of foreign stamps. When it was opened, a large stone lay in it with the following text: '*der Stein der Weisen*' [the stone of wisdom], and everyone had a similar gift. Among other things, I also got a loaf of Swedish pumpernickel bread. I do not know who sent it to me, but I was very happy for it.

On Christmas Day, in the afternoon I began my trip up into the Handeni Mountains. With three Negroes, who carried my clothes, I left Pangani at 2:30 pm, and, after six hours of walking, I reached Kigombe, half way between Tanga and Pangani. It was quite dark when I arrived. But, when the chief in the town (it is called *Jumbe* in Kiswahili) heard that a European was there, he immediately saw to it that I was provided with a house to sleep in. I was, of course, incredibly hungry after the long march and had a large trunk filled with food for the fourteen days I intended to spend before returning home. But, when the house was shown to me, an entire flock of Blacks followed along. All stood and stared at me as if they wanted to eat me. I had to talk to them for about twenty minutes, and, at last, I asked them to leave, because I wanted to eat. Near the hut I was staying in, a wedding was being held with drums, singing and dancing. I was, therefore, worried that they would keep it up until morning and that I probably would not be able to sleep as a result of all the noise. Worst of all, however, the bed was too short. First I let my legs hang out, and, when they became tired, I pulled them in and let my head lay out, just like the lambs in a butcher's wagon. Change is good they say, but the change I experienced by having either my legs or my head hang outside the bed had no good in it. At last I found a solution. I moved the bed against the wall, lay on my back with my legs up on the wall and fell asleep. However, after a couple of hours, I woke up with a feeling in all my limbs as if I had fallen from a four-story building.

In the meantime it had become four in the morning and the moon shone so brightly that one could see as if it were day. I woke up my three Blacks and started out on the way to Tanga, which I reached in the afternoon. No doubt, it is only in the tropics that one has such bright moonlight. Here you can easily read a newspaper by the full moon without straining the eyes. I do not know when a morning has appeared so beautiful to me as this one. We went the entire time along the beach. The sea looked as if a silver sheet had been laid over it. The coconut palms grew close to the beach, and, when the moon stood low, it cast their shadows over the beach and out into the water. The palms moved slightly in the cool, morning wind, and the shadow, outlined sharply on the clear water, looked like some impressive, black fish slowly moving towards land. The beach quickly shifted from flat to large coral cliffs, which, to get around, we had to wade through the water up to our waists. It was not, however, by any means, unpleasant going out in the water so early in the morning, because both it and the air were at ca. 23° Rm.. In Tanga I gained two travelling companions, and, the next morning, we moved out towards Bondei.[14] Tanga will undoubtedly soon be a large town. A railway is being built here, which, at first, will go to the mountain Kilimanjaro. There are a lot of Europeans employed over the entire line as supervisors for the Black workers.

I met several old acquaintances among the Blacks, who, both the men and the women, in recent times, had experienced many strange things they wanted explained. First they wanted to know how it was possible for a man to become as fat as a hippopotamus. They had recently seen a European who had visited Tanga for a couple of days. He was so fat that he had to have two doors opened to come out of the house. When he walked on the

[14] The *Wabondes* are the people of Bondei, near Pangani. They are also known as the Bondei. Resolutely stateless, the Bondei refused to accept a formal leader. The Germans 'solved' this by appointing an alien leader, as did the British initially (Iliffe, 1979:333-334).

No text, but there were no railroads to Pangani or Bagamoyo in the 1890's so it must be Tanga

sand path, he sank in over his shoes and gasped and puffed as if he had ague. Sweat rolled down him, so that all of his clothes stuck to his body. When he came back to the hotel, it took five litres of beer to quench his thirst. His neck was just as thick as his head. Ugh! He was disgusting. (this outburst is very natural for the Blacks, who consider a long neck to be beautiful, and who only find obesity attractive in women). I explained that, when people in Europe became so fat, it was generally because they exercised too little and ate and drank too much. People in certain positions, such as brewers and publicans were nearly always fat, since they drank a lot of beer and did not move very much.

Then others wanted to know how a European woman could be so thin in the waist have such wide hips and such a thick rear. They showed with their fingers against each other how thick she was at the waist, and, by holding their hands out, the width of her hips and thickness of the rear. They could not understand how it could be, because it looked like a piece of meat had been cut off each side of her. I told them that, in Europe, certain circles considered it beautiful to be thin at the waist. To bring this about, since God had not made us that way, they tied themselves up with something we

No text and no date

called a corset, a piece of clothing in which there are sewed steel strips with lacing holes. It was laced up as tightly as possible. It was just like the caravan people, when they tie up their burdens to make them as small as possible. "Yes, but such bound-up women were not married?" Yes, I believe they were. "But they don't become mothers?" Yes they did. "But then they took the corset off?" That I could not answer, because I was not married. "Ah! The poor European children," they said. "But how could it be that they all had such wide hips and rears? Were they stuffed?" Yes they were, with cotton. "Does cotton grow in Europe?" Not much. "So that was maybe why I planted cotton here in Africa, in order to supply the European ladies, so they could be more beautiful?" Yes, of course (general laughter). "Yes, Bwana Malele is an artist. He not only builds houses, but also women of cotton," they said. You must not think that I sit here and make up what I am writing. What I have written here is true, word for word.

We then began to talk about hat decoration. They asked me if it was also considered beautiful to have feathers in hats in Europe. They had seen ladies walk around not only with feathers but also with flowers and European greens in their hats. When I told them that ostrich feathers were considered to be very beautiful as decoration for women's hats, they thought it was very funny. It was, after all, the same taste as the cannibals here, who wear ostrich feathers in their hair when they go raiding after a good piece of meat. All things considered it was agreed that European men were fairly civilized, but the ladies should be considered as savages, because they had such bad taste. That is the Blacks' opinion of taste here. I believe that the European would say the same about the Blacks, when they saw them.

It is hard to imagine what it is like to march in Africa, especially at this hot time of the year. I will try to describe it to you. The trails are about one and a half fod [feet] wide, and the grass grows from eight to ten fod [feet] high. The first day there was no water on the way, and, therefore, we had to take all the water we used from Tanga. Until ten in the morning it was quite pleasant, but, when the sun rose higher and higher, it became hotter and hotter. We could not feel much wind in the high grass, but felt the heat even more. Then the thirst began. The water was lukewarm, but that did not stop us from drinking enormous quantities. The warm water did not, however, really quench our thirst, and, while we trudged on, my thoughts were completely taken up with wonderful, cold drinks. In front of my eyes there floated a couple of large, ice-cold bottles of soda water. A couple of bottles of beer had been opened, and the foam poured out of them. They were so clear that I was close to thinking they were real. Then my companion, who must have been thinking about the same thing, said, "Oh! How I wish we were in Berlin now and sat with a cold glass of beer in front of us." I asked him to be quiet. My mouth watered with just the thought. I just needed to think about beer; I would never become thirsty.

Towards the afternoon we were lucky enough to come to a house, where the owner kindly gave us a couple of glasses of water. My tongue was stuck in my throat by that time, and I had never seen a glass of water look so refreshing as that; and it had such a wonderful taste. It had been kept in a large, clay pot, which the Blacks make. They are quite thin of baked clay, so thin that the water seeps out just like kerosene in a lamp. The seeped water evaporates quickly from the sides, making the pot and the water inside wonderfully cold. Now think about a strong thirst, so strong that one can hardly spit, and then about such a lovely glass of cold water. The best wine in Italy never tasted so good to me. I drank the water in small sips and noticed how my Blacks greedily kept their eyes on the glass to see if there would be anything left for them. I gave them half the glass, which they, in brotherly fashion, shared, and we continued on our way. We now had a three hours march to our camping place. I naturally became thirsty again on the way. I thought only of the cold glass of water, and was irritated with myself that I had been so dumb as to give half of it away. Why had I not drunk it all!

As the sun was about to go down, we reached our camping place. We had marched nine hours that day. You can believe that it was good to remove my boots and to put a pair of Arab sandals on my bare feet. While the Blacks put up the tents and others brought water, we Europeans held council about what we would eat that evening. We had all kinds of foodstuffs with us, which was the reason that it took so long to agree. Finally, however, an entire collection of pots and pans steamed over the fire, and we had a meal which would only possible in Africa after an exhausting march. Our Blacks were, of course, not forgotten but got all that we left, which was so much that they could not eat it all.

The next day we were to go through beautiful Bondei.

In the morning I was up quite early to rouse everyone so that we could leave before the sun rose. Our next camping place was to be at the foot of the Handeni Mountains, which was seven mil [32.5 English miles]. Bondei reminds me somewhat of the lush region around Vejle and Fredericia. All of its residents enjoy a degree of prosperity, but I have never seen such a superstitious people. Whenever the trail went through a village, there was always a little hut, with clay figurines of a man and a woman. The people offered them chickens, goats and sheep, in order to prevent sickness and evil spirits coming into the village. They do not believe in a natural death. If one becomes sick and dies, it is always because he was hexed or the devil had a finger in the game. Then the doctor (witch doctor) is called. He understands how to talk to the spirits. Often he comes to an agreement with them, and the sick person becomes well. If this does not happen, the sick person must die. They live in fear of Satan and evil spirits just like Home Mission in Denmark. When a father talks to his children, it is very much like Wilhelm Beck[15] talking to his congregation, in which the Devil, Hell, sulphur and fire play the main roles.

We went by the English mission Magila,[16] but how much have they accomplished in the twelve years they have been here? About a rifle shot distance from there the people had their idols, and a spirit dance was danced to the beat of drums. They moved around in a circle with screaming and yelling. A sick person sat in the middle, and the evil spirits were supposed to be frightened away by the horrible spectacle. I do not know if she became well or not, since we did not have time to stay. When one thinks about all the want and misery home in England and looks at the wonderful Bondei country in Africa, where the Blacks are well supplied with food and clothing. Here the English throw millions after millions away to help the poor Black, while the poor in London starve and freeze to death, because

[15] Wilhelm Beck (1829–1901) was for many years the leading spokesman of the revivalist movement "Home Mission". As the name indicates Wilhelm Beck and his followers believed there were many "heathens" or non-sincere Christian souls to be saved at home. The Home Mission was therefore just as important as the Foreign Missions.

[16] The English Anglican Mission at Magila remained a centre of controversy during the Arab rebellion against the Germans.

they have neither food nor clothing in the hard winter. Is that what they call charity?

Since Bondei is criss-crossed by small streams, we suffered no want that day. When, in the evening, we approached our camping place at the foot of the Handeni Mountains, we heard a strong roaring sound like the beginning of a storm. Who can describe our joy, when we came through the thick growth and saw a wonderfully clear mountain stream, which roared and rushed down the steep mountain slope. Here, close by the stream's bank, we set up our tents. After we had cooled off, we bathed in its cold, clear water. Such a bath after a hot and dusty day's march in Africa is a pleasure that I cannot describe. How wonderful one feels afterwards, and with what an enormous appetite one sits down to eat. The People of Bondei live by cattle raising and agriculture. The earth has a high clay content. The most wonderful wheat and sugar cane would grow here, if it were not so hot. The most important products are sorghum, maize, sesame, ricin beans, bananas, coconuts along with twenty different kinds of vegetables and rice. They prepare their fields with a hoe and, in an unbelievably short time, can clear a large piece of land of grass and weeds. There is no doubt that they are good farmers. I saw how a large field, with the help of a number of ditches, could be put completely under water, and where the most beautiful rice I have seen in Africa grew. The main ditch was dug out to a mountain brook and walled up with large stones and clay. In the middle of the wall sat a large, wooden plug, which, when they wanted the field under water, was taken out; and the main ditch filled slowly. The small ditches leading from the main ditch out over the field were also equipped with wooden plugs, so that water could be directed exactly as it was wanted.

The next day we began to climb the mountain, which is 4000 fod [feet] high. The entire trail went through the thickest jungle I have seen. The tree trunks were about from 80 to 140 fod [feet] to where the crown began, and what trees! The sight of them would make the heart of any joiner jump. A pink tree had such a fine grain, it could be used for the most beautiful furniture. There were yellow trees and a brown, nut tree, which could make wonderful furniture. On the coffee plantation, where the Chinese work, they had made small chests, chairs and tables of all three types of wood. When one thinks of boards of 80 to 140 fod [feet] of that kind of wood, what wealth there is in such a jungle. If one could only bring it to the coast, but transport is too expensive. A Negro does not carry more than fifty pounds and requires a high wage. When the railway to Tanga, which is close by, is finished, there will be a lumber business here. Foresters from Germany will be engaged to supervise the work, so that the wood is not cut too much. The air in such a jungle is very humid, and the trees drip until eight or nine in the morning. The sun does not shine through, because the trees are too close together. I believe I can dare say that there is a tree every twenty feet. The underbrush consists of clinging plants, small bushes and the most wonderful ferns, much larger than those in Denmark.

The soil is everywhere red, rich clay, and, in the morning, when everything is so wet, one is constantly falling, because the path is so slippery. But the mountain is not climbed at once. There is first a little hill, then down into a valley, up a larger hill and down again, and so on, until one has come up to 4000 fod [feet]. The entire time, a foaming mountain stream rushed past us, between the stones and tree trunks with wonderfully cold water. I believe it was better than the water in Blegkilde in Aalborg,[17] or perhaps it tasted so good, because of the exhausting walk tour up the mountain. We reached the coffee plantation we were to visit at ten in the morning, where we were heartily received by the manager. I see that I am running out of paper and can, therefore, not write more at this time about my trip, but I promise you that I will write a long letter about Bondei and its people. This is only a short description. I cannot, despite my best wishes, make it longer, since the letter must be sent this evening at five and I must send it a mil [4.65 English miles] over land.

I will write to Mrs. Gay tomorrow. It will go with the French post to Marseille. Have Hjalmar write me when he leaves. I am now in correspondence with a friend in Columbia, South America. If Hjalmar later wishes to learn coffee cultivation, I may be able to get him a position there.

Friendly greetings and a happy New Year to all
From your affectionate brother
Chr. Lautherborn
Don't forget to send a photograph of all of you.

[17] When Aalborg in the 1850s got a public water supply one of the main sources were Blegkilde ("kilde" means spring). The water here was considered to be tasty as well as healthy.

CHAPTER EIGHT

KIKOGWE AND MWERA: COTTON AND COFFEE

Nguelo, 15th March 1895

Dear Sister and Brother-in-law
As you can see, I am no longer in Kikogwe. On the first of February I received a telegram from the company's director saying that I should go up to the Handeni Mountains and take over the management of the two coffee plantations, which the company has up there. I believe I described a trip up to the named mountains along with a couple of good friends. The reason that I have not written in such a long time is that the company is really giving me too much to do. Here both managers have left at the same time, and Lautherborn is, of course, chosen as the one who shall work for two. I said to the director, when I was in Tanga, that I actually had enough to see to with the one plantation. I thought that I was too long in service with the DOAG to carry out such hard work, when they had younger men who would benefit from standing on their own feet. He, however, thought that the harvest was ready and he could not turn the plantation over to a younger man. The company had spent a million on the two plantations in the three years they had existed, and, if the coffee harvest, as a result of poor treatment, were spoiled, hell would be loose in Berlin. No, there could be no question of anyone else; I had to go up there. Well, I thought like the Americans: "Life is duty, and let us try to do our duty the best we can; then we cannot do any more."

After a difficult march I arrived here on the 6th of February. There is really wonderful coffee here, quite incomparable. The branches are so full of fruit that they bend down to the ground. But now the worst of all, the harvest had to begin in about two months. There is no building built for cleaning and storing the beans. There had to be a building where the coffee could be pulped, fermented and washed and one where it could be sorted, rinsed and dried. The character who was here is, naturally, sitting in Berlin now, drinking cold beer and laughing over getting away at a time when there is so much to do. In six months he will return, thick and fat, grinning all over his face, and ask me if I was strong and healthy after my rest in the wonderful air up here. But he will not go unrewarded. I will find some way or other to pay him back. As a matter of fact the climate up here is delightful, that has to be admitted. The air is rarely over 20 degrees Rm, while the mornings and evenings are between 10 and 15 degrees Rm. When one has been so long down on the coast in the heat, it is a virtual paradise to come up here. Recently we bathed in 13 degrees water, but I had to pay dearly

'My coffee-plantation'

for it. I got dysentery from it and had to stay in bed for six days. I stood up so thin and exhausted that I could not even go across the floor alone. I am now well again and shall be careful about bathing in the streams here, because it is too cold for me.

On the plantation here about eight different languages are spoken. We have Chinese, Javanese, Tamils, Malayan, Negroes, Germans, English and French.[1] At times there reigns a true, Babylonian confusion, when I give orders about work and schedules in the morning. Often one really does not know what language one should use. It is a good thing that most understand Swaheli, so one has, despite everything, a main language. Most of the Chinese, however, only know their own language, and, with them, it is quite difficult to make oneself understood. There is, for example, my gardener. He is a very important person for me, since I set a high price on all kinds of vegetables and would like to give him instructions and help him

[1] In 1902 DOAG contracted some 462 Chinese and Javanese indentered labourers from Singapore to work on its coffee and tobacco plantations in Usambara. Abuses against these workers led to refusals to continue this practice (Sunseri 2002: 55).

with the garden. But he only understands his native language. It must be funny to see, when we two try to make ourselves understood to each other. It probably looks like two deaf-mutes, because we speak only by sign. However, we understand each other, and I think the Chinaman is happy when I come to speak to him; but I could not say for certain.

Before I left Kikogwe, Peer Scavenius came to me. He is a son of the Minister of Culture. He is now working in Kikogwe as a trainee and wants to learn to be a planter. But why has he come to Africa? His father has, after all, several estates in Denmark, and is not it his uncle who lives on Voergaard?[2] He is a very nice person, more than that I cannot say. I do not believe he was used to doing anything at home. I turned over my books to him when I left. He wrote to me recently that it was going very well with the bookkeeping and that he had figured out everything. But his handwriting is a horrible scribble; it does not look good in the books. But it gets worse. Is he not married? There are various rumours here that he has left his wife, but that is probably not true. We both knew a number of farmers and farms. He has been in Jutland a great deal. We had lively conversations the last days I was in Kikogwe.

There is another Dane here, a Hr. Kuhfahl from Fredericia. He says that he knows Cathinka, but the Devil knows if that is true or not because he lies as fast as a horse can run. According to him, he has been all over the world, took part in the war down in Argentina, gained the rank of general, was decorated with three medals, later became a captain with his own ship, which was, however, taken from him, because he was too brave and became involved in a fight, when he should have been neutral. He was a tobacco planter on Sumatra for several years, and later he became a merchant marine captain and earned 12,000 a year. But he smuggled too much on the Chinese coast and, therefore, lost his position. He then went into Chinese service, was later a merchant and much more. The man is thirty-five years old, but I have figured out for him that, since his twentieth year, he has, ac-

---

[2] Peer Scavenius (born 1866), son of Jacob Scavenius, among other things Minister of Culture and Education 1880–1890. The family was one of the richest and most influential families in Denmark in the 19th and early 20th centuries, belonging to the Danish nobility and owning several large estates. Voergaard is a manor in North Jutland that belonged to Peer Scavenius' paternal uncle. Peer Scavenius seems to have been 'the black sheep' in the family.

He participated in *Frilandsexpeditionen* [the Free Land expedition]. The expedition was an offspring from a German-Austrian Movement. The movement and expedition aimed to found an ideal society which was neither liberal nor socialist, but tried to combine the best from the two societal systems. Peer Scavenius wrote a book in 1897 where he explains the ideology of the movement and also tells the story of why the society he and others in 1894 wanted to establish in an area behind Lamu failed (*Frilandsexpeditionen. Dens Tilblivelse, Forløb og Undergang*, Kbh. 1897). He married in 1892 and divorced in 1907, so the rumour about him having left his wife in 1894, which had reached even CL, was probably right.

cording to his own description, spent fifty years in different positions. He did not like that. He has tried to become a close friend, but he has not succeeded, because I have no trust in him. He, for example, loaned me a good number of S. Schandorph's stories,[3] but I found them very boring and was going to send them back to him. He was thrown out of the plantation where he kept books. What the reason was I do not know, because I do not associate with anyone on that plantation. Now he wants to start a business in Tanga, but I do not think he has the money for it. However, ask Cathinka, when you see her, if she knows him.

Now I can go home in seven months. I don't think I will though, because it would be in winter, and it would be very cold. It will be better if I wait until next summer, then I can get out and see something. I don't really need to ask how you are doing. I, thank God, am fine. I have a lot to do from morning until night, but it makes the time go; and one is happy at the end of the day that one has accomplished something.

Greetings to all the children, your husband and yourself from your affectionate brother
Chr Lautherborn

Deutsch Ostafrikanische Gesellschaft
Mwera, 25th December 1895

Dear Sister and Brother-in-law
Thank you very much for the almanacs. They reached me exactly on the afternoon of the 24th of December. I had a little fever and lay in my easy chair resting, when the Negro came with them. It was my Christmas treat this year to read them and the papers that were also included. Since I did not feel really well, I stayed home. After all, what pleasure is it to go to Pangani Christmas Eve. It is true that one is together with good comrades, but, on such an evening, one has to eat and drink well. Since the Africans are neither moderation nor temperance people, one usually has a hangover Christmas day; and that is not much of a pleasure here in the tropics. *Familie Almanakken* [The Family Almanac][4] was the first I put my hands on. I leafed through it and looked at the pictures. Suddenly a street or a picture

---

[3] Sophus Schandorph (1836–1901), poet and novelist. His first poems where written in a romanticist style. He later wrote short stories (humoristiske folkelivssskildringer). He belonged to the progressive circles centering around Georg Brandes in Copenhagen in the second half of the 1890's, but it is most probably some of his earlier, now forgotten short stories, that CL had borrowed from Hr. Kufahl.

[4] The name 'almanac' lingered on for a certain type of annual publication, although in a strict sense it was not an almanac but a booklet with novelettes etc. These 'almanacs' were popular in the 19th century. They were normally printed for sale around Christmas.

of someone will strike me. Yes, he looks good as a soldier, but I think he could have chosen a more peaceful face for Christmas. The picture of the farmers from the islands and Jutland is really good. One thing, however, is seen immediately, nearly all of them have no teeth. What could be the reason that the core of the Danish people have such bad teeth? That is something that I would really like to know.

Holger Drachmann's[5] poem about Skagen amused me greatly, one sees, however, that he has no interest in farming. If he had, it would have been quite impossible for him to write so beautifully about lyme grass, dunes and drifting, flying sand dunes. I can understand writing about the islands and the Vejle area as places where one would like to stay. But Skagen with its sharp wind, sand and lyme grass, I don't understand how one could swoon over that. Yes, perhaps for a day or two, that I can understand, but no more, so away to where one can see green trees. If I meet Holger Drachmann in the summer in Skagen, I will advise him to go to the Sahara, since, after all, there is plenty of sand. Well, he appears to also feel that only few share his taste, for he says at the end, "I am, therefore, a camel." One cannot blame Holger Drachmann for that, but he does seem to have quite a strong fantasy. But back to the agriculture pictures again. When I was home, one didn't see any homespun coats and hats, but here in the picture, one sees, by God, the Danish farmer as he looked when I was a child. I just do not understand why only old people were chosen. There are only two, young girls in the entire collection, and they are from Sjælland. I know one of the farmers from Jutland. He is the second from the left and the third from the top. He is sitting and squinting with his eyes. I have been with him many times; I just cannot remember where. I do not think Chr. Winther's 'Pale Rose-coloured'[6] is as beautiful as one would believe from the description, but perhaps it is not a good picture of her. It is, however, possibly a good likeness, because Chr. Winther was, after all, at an age, when he married, at which one sees more of the interior than the exterior. The picture "A Proposal" is the best in the 'Almanac', but the poem by Chr. Winter does not fit it, because it states, 'You lift your cold eyes blue towards my wet eyes'. Since he is sitting on the bench, and she is standing up it should be, 'You lower your eyes blue....' Besides I don't understand why the student is in such a rush to marry. He could easily wait a couple of years more I think. He is not so old.

I have set up the map of Denmark in my living room, and will now study

[5] Holger Drachmann, (1846–1908) belonged to the colony of artists, mainly painters, who from the 1880s on lived in Skagen during the summer. The painters belonged to the impressionist school and poets like Holger Drachman belonged to the modern breakthrough in Danish literature 1870–1900.

[6] Christian Winther (1796–1876), Danish poet, romanticist, in his days highly esteemed. Regarding style and themes he may be compared to Heinrich Heine and Lord Byron.

it. It was my intention to take a little trip to India before I went home, but now, after seeing the map of Denmark, I have changed my mind. In all honesty it is disgraceful that I now have been to all four corners of the world, but Copenhagen, the capital of Denmark, I, a Dane, do not know. That must change. I was happy to read in *Vendsyssel Tidende* the answer to Vilhelm Beck concerning his Home Mission nonsense. It is really sad that the otherwise so wise Danes would allow their old God, who is good, and whose mercy is eternal, to be swindled from them, and, instead, believe in Vilhelm Beck's God, a God with fire and sulphur roaring out of his throat, who burns and tortures all of mankind for small faults just like the cannibals here. I know that what Vilhelm Beck says is a lie. Mankind are the children of God. He is a good and loving father, is patient with our small faults; this we learned when we went to school. Now Vilhelm Beck will suddenly give us a stepfather, who has less sympathy for us than a Sahara lion for its prey. He can keep that God for himself. I hope to our old God that Vilhelm Beck's new religion will not grow too strong. It was really sad to see, the last time I was home, I think it was in Sæby [small town close to Aalborg and Hjørring], such a fire and brimstone priest hold an assembly, and, when the people came out of the church, how fearful and miserable they looked. There was nothing of the satisfaction and joy seen in a congregation, when they come from a church, in which a good priest has delivered God's word to them in a good and loving manner. No, the step was heavy; the expression was dark and fearful, as if all will to live had been crushed. It was not any wonder. If they believed what was told them, that they only lived here on earth so that their soul could be tortured in hell after death, there was good reason to look dark and miserable. I will only hope for my homeland that enlightenment moves forward faster than Home Mission. Otherwise it will look bad for Denmark in the future.

Yes, now I am here down on the coast again, and had begun to think a little about going home for Christmas. It was, however, the severe cold in Denmark that held me back. I sent you a sack of coffee for Christmas; it has probably arrived. I have four beautiful panther skins and four mountain monkey skins. I am sending them to Cathinka. She will keep four and send the other four to you. When they are tanned, they are good for laying on the bed or the sofa in the living room. They keep the feet wonderfully warm. Yes, I am now looking forward to going home. There are only two more months, and so I will say farewell to Africa for a time. Yes, perhaps for good. I still have no contract with the company and do not know if they will accept my requirements, but I believe that they will. I do not believe that I have given them grounds to be dissatisfied with me. I believe that I have written you before about Bondei, the fertile land that lies at the foot of Mount Usambara where I was on the coffee plantation. I also believe that I told you one time about the superstition that rules here, which even the English mission cannot eliminate.

No text and no date—The Pangani river close to Kikogwe

No text and no date—The Pangani river close to Kikogwe

The Bondei people are of medium size, quite well-built and more slender than stocky. Among the young, one finds many that are good looking. The old, on the other hand, are usually horribly ugly. They are timid and dependent, but good-hearted people. At the birth of a child, they have many different customs. From the beginning of labour until the birth, the father must not eat anything. The mother and child must not leave the hut for six days, and, during that time, the mother must not eat meat. Only on the 7th day may the father see the child. He then gives it a name, brings a basket with clothes and bananas and sends his father-in-law a goat. With regards to names, there is a great confusion in Bondei. Every child is named after his grandparents but also has many nicknames. When, however, he first becomes a father, there takes place an odd name change. If he, for example, gets a son, who he calls Mbuju, he is called Sembuju and the mother Mambuju. If Mbuju marries and calls his first born Ponda, he is called Sepondo. If Sepondo marries a second wife (in Africa there is nothing new in a man having two wives), who gives him a son, who is called Kalata, Sepondo continues to be Sepondo, however his second wife changes her name to Makalata. Only the first born's name has influence on the parents' name; all of the other children's names have none. Child murder is frequent. Twins and children whose upper front teeth grow first are killed because they are bad luck.

As soon as the boys are big enough, they can herd cattle, but only for a couple of years. They then have to do field work. Bondei girls are also taught at a young age to take part in all housework. At the age of twelve they can often perform the same work as a grown woman. Boys' games consist mostly of making bows and arrows and sling shots, while the girls play with dolls they make of clay or coconuts. They rock the little doll baby on their arms or carry it in a cloth on their backs, while they very seriously, pound meal in a wooden mortar in order to feed the little one. Boys or girls who are about twelve no longer sleep in their parents' huts, but the girls get their own hut, and the boys get their own hut to sleep in. I was told that the children were thus separated for the sake of decency. As soon as the boys can do the work of an adult, their tribe's mark is cut into the skin by the witch doctor, a kind of tattoo. When this is done, the boys and the witch doctor go into the forest, where they build small huts and remain there over a month. During this time they are taught all kinds of magic and must not wear any clothes. Only about the hips there is bound a piece of bark. The body is covered with lime or ashes. False beards are made of goatskin, which they tie on themselves. They have bands around their legs with small bells and carry bells in their hands. In this outfit there is constant dancing and singing, and only what is absolutely necessary to maintain life is eaten. The food is put into the forest by their families, close to the place where they dance. This practice must be gone through by all who want to be considered part of the people, if not his children are considered illegitimate and are killed. It is the same with the girls, when they are declared adults by their parents. First the tribal mark is cut into their skin. Thereafter, quite naked, they go into the forest with an old woman, where they stay for eight days. They can, however, during that time, return to the village to perform some errand or other, but always naked. The closing dance takes place in the village, and all the young people in the area take part. The young girl sits naked on her mother's outstretched legs and is painted on the face and body with white paint. Later she gets a glowing piece of coal in her hand and must run as fast as she can with it through the town. The dance lasts for two days, and all who have legs dance and drink palm wine.

If a young man wants to marry, he sends his chosen one's father a little present consisting of palm wine, a basket of rice or the like. If it is accepted, the two are considered to be betrothed. A few months later the bridal price, consisting of calves, clothes and money, are determined and paid. Thereupon the wedding takes place. The bride and groom go, one morning, together with friends, to the hut selected for them. Here they take their places on two beds standing next to each other. They must stay in that position for four or five days without nourishment other than a little milk to drink. They must not leave the hut, in which a large fire burns. On the last day a meal is brought, and the young wife eats chicken for the first time in her life. They then sit outside the hut. Corn, as a symbol of fertility, is sprinkled over their heads, and, with the firing of shotguns and dancing, the celebration ends. Polygamy is allowed but not very widespread. On the wife's shoulders rests the house and a great deal of the field work. She must fetch water in stoneware, often from long distances, chop wood, and grind corn. The corn is ground by the following method: a large, wide stone is positioned so that one end is higher than the other. The stone has a small depression in the middle. The corn is spread out in the depression and ground with a smaller stone until it becomes flour, exactly as painters at home grind oil colours. She must also cook the food. That is not, however, associated with so much difficulty as at home, because there is rarely cooked more than one meal per day, otherwise the husband and children roast potatoes, rice or the like.

Here on the coast, among the Swahilis, it is done another way. When a young girl is mature, (she is called mature when she reaches puberty), it is celebrated with singing and dancing. The young girl goes through the town without saying a word, and, with lowered eyes, she puts out her hand to passers-by. If she gets a gift of money, which happens frequently, she speaks, if she gets none, she remains mute. She then has seven days of training with an old woman. Part of it is about the obligations of a married woman in regards to faithfulness and obedience, and part of it is practical in the form of obscene dancing in which the old woman plays the man. The girl must also learn massage, for it is a Swahili wife's obligation to give her husband a massage every day, if he wishes it.

A young Swahili soon looks for a wife. Since there is no *purdah* [seclusion], and the girls go without veils, one can quickly choose a future bride according to one's taste. When that is done, he sends a negotiator to the girl's father. If he gives his approval, which is usually the case, the bridal price is set. This is seldom under 100 Rp (A Rp is 1 kr 10 øre). Thereafter the wedding takes place according to Moslem custom with a banquet and dancing. The expenses for this are paid by both sides. The thickly made-up bride is escorted to the bridegroom, and a white sheet, which is publicly inspected, is laid over the bed. The following activity takes place in the presence of a couple of old men and women. If the sheet becomes, thereby, bloody, the old women in the bridal chamber issue a penetrating cry of pleasure which is enthusiastically answered by those waiting outside. If the opposite is the case, there is no cry of pleasure, and the bride is overwhelmed with insults when she comes outside, because this is considered a great shame for the parents as well as for the bridegroom. The latter also acts as if he were very much in despair, but soon calms down. To avoid all the unpleasantness the guilty girl usually cuts herself on her leg. It also sometimes happens that the old women are good enough to drop a little chicken blood on the sheet and then give the cry of pleasure. The bloody sheet is then laid on a plate and carried around the town, and everyone has the right to inspect it closely.

If a bride's father is dead, the bridal price is paid to her relatives. If she has none, it is paid to herself. Only if the bride is completely independent and wealthy and the bridegroom poor can the bridal price be waived. If a Swahili wants to make his lover his legal wife, he must first declare her to be free and, for a short period, have her live in another house. He can then pay the bridal price to her and marry her.

Few Swahilis have several legal wives. However many have, besides their wives, several *surias* (concubines). These are directly bought, but every man has the right to take one of his field slaves that he likes as his *suria*. This law is very hard, because it also includes his married slaves, who, in this way, are divorced from their husbands. It is, however, only the Arabs who are hardhearted enough to make use of this law. In general wives as well as *surias* of Swahilis are treated well, as they are not in possession of the jealousy with which Arabs unceasingly ruin their women. By tradition and upraising the wife has acquired the conviction that the husband is her unlimited lord and master.

With the easy view of life of these people, however, this principle is not too strictly maintained. True, the man usually gives his wife, but also his lover, a couple of good blows with his walking stick. This, however, is more for the sake of form, and rarely develops into mistreatment. The beatings are always received with patience as evidence of the man's love. There was, therefore, great bitterness here, when an old Swahili beat his young wife in order to maintain the good old custom, and she, instead of accepting it as proper, gave him a couple of strong blows and threw him out of the house. This example shows that the weaker sex's humility and submissiveness, even in East Africa, is not to be depended on. Soon after the wedding the young husband will spread around the house money or other things of value to see if his wife will steal. If this proves to be the case, or if he catches her in adultery, he has the right to send her back to her father. The father must repay the bridal money; and the marriage is dissolved. A husband, without definite grounds, just because he is tired of her, can send his wife away. He does not, however get the bridal money back, but on the contrary, must pay her a certain sum of money. If a wife does not want to stay with her husband any longer, she tries to get her father to pay the bridal money back. If he does this, there is no obstacle in the way of their being divorced, and she can marry again, if she wishes.

When a child is born, it is immediately given a cap, which it must wear until the hair grows. Then one usually gives it a thin, corn gruel to eat, since it is assumed that, as long as the child is borne by the mother, it gets nothing to eat and, therefore, screams because of hunger. If the screaming does not stop, however, it is given sweet palm wine with sugar until it sleeps. The children drink mother's milk for a long time, even when they are old enough to eat meat and rice. I know several people here who clearly remember drinking mother's milk. The children are divided into different groups. A child, if the mother and father are free people, is called *mtoto wa watu* and stands on top of the list but with, of course, about the same rank as others in his class. A child born into a marriage of two slaves is called *mzalia,* who is born in the country a slave. Much higher than that, nearly on the same level as *mtoto wa watu*, is the child born to a free Swahili man and his concubine. It is called *mtoto wa suria*. These are all the legitimate children. All others are illegitimate and are called *watoto wa haram*. The last are found everywhere. They are often the children of a free man and another man's slave. This is judged fairly mildly if she is not a *suria* of her owner, and the father can buy the child and later make him a free man.

Illegitimate children of a free man and a free girl are immediately killed. All in all child murder is quite frequent and is carried out by choking or drowning. If a woman does not want any children, she first tries to expel the foetus by eating a large amount of pepper. If that doesn't help, the newborn is simply killed. This is seen to be quite ordinary, not as anything wrong, and people here cannot understand that we Europeans become so upset over such a natural event. The small child's life hovers in great danger when it gets teeth. If the lower teeth do not grow first, but on the contrary, the upper teeth or molars, it is immediately strangled by laying hands on the throat. A child who is born with teeth is also killed. This still takes place today, even though the Germans do everything to prevent it. The people, however, know how Europeans judge such a procedure, so they are very careful that no one learns anything about it. If the child has survived this dangerous period, it grows very sturdy and learns to walk and talk more quickly than European children. It acts much more independent than European children. Its toys are a small doll, a canoe and a bow and arrow.

That was a little about the domestic conditions found among the *Bondei* and *the Swahilis*, but, before I end the letter, I will talk a little about the Arabs as they are here. I will, in that way, get good use out of Christmas Day, and perhaps it will please my countrymen to read it. First, however, I have to eat and smoke a pipe.

I believe that I have, in my earlier letters, made it clear that the Arabs are not my friends. First because their character is not open and honest. Second because they almost never tell the truth, and third because most of them are cowardly and cruel. The fact that they had won such a firm foothold here they have their Moslem religion to thank for, with its polygamy, slavery and, finally, religious ceremonies it is, of course, a religion excellently suited to the Negro, and the Arabs here have completely controlled him with religion. The Arabs naturally have many slaves and treat those who work in the field and take care of his house quite well. His *surias* (concubines or harem wives), on the other hand, have a truly heavy burden to bear, which comes very close to the sad picture painted by Europeans of slavery. It is also, as I wrote before, very hard that the Arab takes any of his field slaves as his *suria*, even though she may have been long-married to one of his slaves. In the last case the often very old Arab watches, with insane jealousy, over every step the unhappy creature takes. His revenge is fearful if he catches her together with her former husband.

As an example I will tell you about an incident in one of the Arabs' harems in Tanga. He himself died here during the rebellion. He was very good and mild to his field slaves and fed them well and plentifully. For him they were like useful animals, which one always, as much as possible, protects and cares for. On the other hand, for the 70 year-old ancient, his *surias* were only toys, and to them he was hard and cruel. One day he discovered his favourite daughter's mother in a conversation with her former husband. Even though she had been his *suria* for several years, and even though the little daughter had unmistakably inherited his Arabic features, and he had no reason to suspect his *suria's* faithlessness, he decided to kill her. He behaved with her the entire day as if nothing was amiss. At nightfall, with the help of two young slaves, he had her bound to her bed, so that her head hung over the edge and lit a choking fire under her mouth to prevent her from screaming. He whipped her with a strip of hippopotamus skin until she died. She was buried. The next morning the Arab, with smiling face, went to his little daughter and told her that her mother had moved away. After a short time, however, the rumour of the murder spread, and the Arab was called to the sultan in Zanzibar. It was not, however, so strict as it is today. After a couple of days he was released and returned to Tanga. Soon afterwards he stabbed another *suria*, who had an improper relationship with his own son (Fereschi bin Abedi). But he met his fate at last. One of his *surias* poisoned him and three days later married a Black. Of course it could not be proved that she had poisoned him, as the law was then in a very bad state.

Very characteristic for the condition slaves live in is an incident that took place a few years ago. At that time there was a rumour, generally believed, that a slave who came onboard an English warship was immediately free. When, one day, a British corvette anchored in the harbour near Tanga, a large number of slaves ran away and came onboard during the night. Naturally the rumour was wrong, and, the next day, all of the slaves were set ashore by the English. Of the field slaves only a few ran away, and they were the most lazy, worthless creatures so their master would not have been very sorry if they had gone. Of the house slaves a few more ran away, but without exception, all of the Arabs' *surias* ran away and left quite a few children behind. The wailing must have been great among the Arabs, who, in the middle of the night, saw their harems robbed of their best ornaments. Of the Swahilis only a few *surias* ran away and only those who were in family with the other run-aways. One can thereby see that it is only the Arabs' *surias* who are badly off, and I am quite convinced that the day all slaves are free there will not be one, single *suria* with the Arabs. They will leave immediately. On the other hand many of his field workers will stay with him of their own free will.

The Arabs live in stone houses with flat roofs. They are nearly without light and air, that is doors and windows; they are just as sinister and dark as his own mind. Previously the Arabs were the highest nobility in the land, a class of people who were far above the natives and whose influence and commands they passively obeyed. It is difficult to say what they have done or what they do, but one thing is certain; they have never worked. With the exception of a few, poor Arabs, who try to make a living as traders, they are nearly all farmers, who, with the help of a number of slaves, cultivate the land. On the side they have a little trade and occasionally go with a caravan into the interior. They are, of course, all strict and fanatical Moslems. From the height upon which they once stood they are now sunk significantly lower. For the present rulers in East Africa (Europeans), the Arab is not higher than the Black, and his political role for that reason, has been played out. It is not so seldom now that the Arabs take a great beating from the Blacks. Even his *surias* now have the courage to complain if he does not treat them properly. It is also very good for the Blacks that the Koran is no longer allowed to be the judge. Before the Arabs have always managed to get out of it what would suit their case, so that they were always right and the Blacks always wrong. The Germans, however, go slowly and surely forward. It will not be too many years before all the bad practices that now exist in East Africa cease. Well, one cannot say that I have been lazy today, but now I am tired of writing. I wish you all a good and happy New Year.

Yours affectionately
Chr Lautherborn

[A number of pages have been left out. In the summer of 1896 Lautherborn visited his family home in Denmark. One letter from CL to Caroline Carlsen about an outing with CL's old parents has been omitted, as well as letters written on the voyage back to East Africa telling of Naples and Rome—these pages add nothing regarding CL's African experiences]

Kikogwe, 27th June 1897

Dear Parents,
Father's last letter of 16th May and the previous letter, in which Axel had enclosed a letter, I have received with pleasure. Since today is Sunday, and the busy time of planting is over with, I am now sitting by the open window, from which I have a beautiful view over land and sea, writing you. I am sorry for Jörgensens and especially for Miss Alma over their loss of Mr. Petersen. I can fully sympathize, because a good friend of mine has just drowned. He was the postal secretary and was going to go onboard a coastal steamer, which lay at anchor a little out at sea, in order to fetch the post. On that day we had a very high sea. The boat turned over, and three people were lost. My friend, a Portuguese and a Negro, was one of those types of people one seldom meets. I do not believe that he had any external advantages, but he was open and honest, courageous, truthful and loyal. These are traits one seldom meets. The same composure, however, with which he had fulfilled his difficult duties every day, accompanied him also to his death. He neither cried nor screamed, but quietly said "I cannot swim." He knew, however, that he died in fulfillment of his duty and went, therefore, calmly to his death, because he was conscientious. I found his body the next day. It had washed up close to my house. He had come in, I believe, to say goodbye to me. The next day he was buried in Pangani. One could best see how much he was liked by the many tears that fell at his grave.

As I said before, the busy time of planting is over with. However we still have enough to do with cleaning and keeping the earth loose around the young trees. I am, you see, alone here. My assistant is on holiday in Europe, as I was last year. I did get an assistant in his place, but he had to go up in the mountains again to help with the harvest. On the other plantation, Mwera,[7] I have a European, but I have to go up there a couple of

[7] Mwera and Kikogwe are listed as the great sisal-hemp plantations of DOAG, both located South of Pangani, in *The Handbook* 1920:124. "Kigogwe also makes an immaculate

No text and no date, but the photo must be from one of CL's plantations, most probably Kikogwe

times a week to see if the work is done right. I also have here 200 Blacks to see to. So there are not many free minutes per day left to me. It is first in the evening that I can take a comfortable rest in my easy chair and read some good book or other. But work is a pleasure, when one has such good results; and everything grows so well as the coffee here right now. Next month I will send my first harvest here from Kikogwe to Hamburg. It is not much, 100 pounds at most. Three years ago I planted 100 Liberian coffee plants. They are now trees of more than six feet high and sit full of fruit. It is, of course, an important event, not for the sake of the money it brings in, but one, thereby, learns what the price is and can estimate what the plantation will bring in when we have 400,000 trees.

---

impression, especially the crops at Mwera which are beautiful, and it goes to show that it does make a difference which kind of ground sisal is planted in." Kikogwe and Mwera have four factories which together can produce 24 bales or 500 lbs [of sisal] daily. Unfortunately the labour shortages were so problematic that only 3 of the factories were able to function. As a result Kikogwe did not produce a full 1000 tons when in theory it should have delivered 1350 tons...(Zanzibar National Archives AL2/60/39, quoted from Steve Fabian's notes).

Kikogwe, 1st September 1897

Dear Sister!
Your letter of the 18th of July reached me here in good condition, and I see, to my alarm, that you expect number twelve. That the number in question has made its appearance in the form of a little girl I have learned from Hjalmar [CL's nephew, Caroline and Vilhelm Carlsen's oldest son], who wrote to me some time ago. It was his opinion that he would spend his old age among an entire flock of old maids, since he could not see where he would get brothers-in-law for all those girls especially in these times, when no one marries. There is, after all, some truth in that. But I will now write to him that I will help with the difficult work. If nothing else works, I will take some of them down to Africa with me and marry them to some millionaires I know here or maybe to a couple of Arab chiefs, so that they can be carried around in gold chairs all day. I am now also worried about how it will go with all those girls, but, if we all help each other, we will probably get them all married.

Yes! I can well believe that mother does not care for Margerita, but that is only because she has not seen her. She would then, no doubt, have taken a better view of her than I. Yes! For Hjalmar it is good that he is in Switzerland. He wrote and told me how it was going for him. It was, of course, not very pleasant, but it worked out; and he now feels that he is all right and is thinking about going to Paris. Well, I can visit him there in two years, if nothing comes in the way.

I have just come home from our tennis court in Pangani, and believe me, it is beautiful. One side is next to the river, but far enough away that the balls cannot reach it. The other three sides are surrounded by palm trees 60 to 70 feet high, which are wonderful shade from the burning sun. To fetch the balls we have a flock of small Negroes, who, with shouting and laughter, bring them back. Even though we rarely begin to play before 4.30, it is a warm affair. The sweat rolls down the back so that all clothing is soon soaked. That is why I avoid playing too much. When, however, we have ladies here, one has to do one's duty for their sake. I said ladies and mean by that the Prefect's wife, a beautiful brunette of 200 pund [pounds] and her adopted daughter, a Negresse of fourteen years, slender with small, lilliput feet, who speaks very good Viennese German.

The entire family is from Vienna, which one immediately detects from their language. The Prefect is somewhat of a nonentity and is, therefore, not discussed here. When I have Miss Lily (that is the name of the Negro girl) on my side against the Prefect and one of his officers, they are always beaten. When I play to her, I always hit my ball so that she has to run a great deal to come up to it. She then sweats so that it is a pleasure to see. She is then upset and does not say much, and I enjoy myself in silence. When we are assembled in the evening, however, she has forgotten every-

'Lay down of the foundation stone for the sugar-factory 27th January 1899'

thing from the courts and is lively and friendly. Then my consciousness nags me a little over giving her such poor balls, but do you think it is a sin to give a fat woman something to run after?

As you can see, I began this letter in October but have first managed to finish it now. Thanks for the books. 'Peters Jul'[8] is perhaps written more for children. None the less it put me back on Søndergade, where Mother's pastry sack hung over the stove; and where, as far as I can remember, we ate rice pudding and meat balls with red cabbage Christmas Eve, because we had no money for a goose then. 'Blæksprutten'[9] was excellent. It greatly amused me. There is, of course, a great deal in it that I do not understand, since I have only a poor knowledge of conditions in Copenhagen. It was witty though.

[8] Peters Jul [Peter's Christmas], a nursery rhyme first published 1866. A new edition with coloured pictures came out in 1889—it is probably this edition CL's family sent him. Peters Jul telling about preparations etc. for Christmas, seen through the eyes of the child Peter, codified what a true Danish Christmas should be like for generations to come.

[9] *Blæksprutten*, an annual satirical magazine, published around Christmas, with texts and cartoons, mainly comments on Danish politics. First published in 1889.

I have now, thank God, come so far that I have time to write. I had very little of it before, when my assistant was in Europe. Now that he is back, however, it goes well. Emil de Lemas writes and sends me your greetings from Hjørring.

Mwera, 4th April 1898

Dear Parents
Thank you very much for the letters. It was funny, all three arrived at once; I then had something to read. Mother's letter was, however, the best. I studied it for two days, until everything was clear to me. It looks like she must have had a wonderful birthday. That Cathinka had plenty to do answering the door bell and receiving presents is obvious, when one has become as popular as mother. I am very happy to know that you are in such high-course in your old age and are doing so well. I hope it all continues for a long time. Yes, Christmas, good old Christmas, we have also, naturally, celebrated it here. But what is Christmas in a heat of 26 degrees Rm. No, it is only a white Christmas that I would like to experience once again. We were, of course, in Pangani, where we had a Christmas tree; and where a variety

No text and no date, but it may be the Christmas party mentioned in the letter 4th April, 1898

of beautiful presents were exchanged. We also sang 'O *Tannenbaum! O Tannenbaum! Wie grün sind deine Blätter.*' Since we did not have a goose, we had a really good roast duck with red cabbage, a good soup, toasted bread with caviar and several good things, which were very enjoyable. The next morning, Christmas day, my carriage, which I will later send you a picture of, was on the other side of the river.

I had invited all of the gentlemen out to me. We now drove and rode out to the Mwera plantation. Here we had a really good table laid out, all cooked according to Mrs. Nimb's Cookbook. Lieutenant Tank gave a speech, in which he especially emphasized the good food he had gotten and thought that he could not eat better at *Kempinski* in Berlin.[10] He said that Lautherborn here was the man for good food, "Er lebe hoch" [long live] etc. I, naturally, thanked him and said, "It is not my fault that the food is good but Mrs. Nimb's, whose cookbook I have used." There was then a speech for Mrs. Nimb, and all agreed that a telegram should be sent to her. But, since I did not know where she lived, nothing came out of it. As the next day was Sunday, we discussed the best way to use it. I suggested that we should ride to Makerare, a place a good hour's ride from here, 500 fod [feet] above sea level. This was accepted with cheers, and things got lively. Negroes were sent after donkeys. I went to the kitchen and put the cook to work making food for the next day, and, the next morning, at four, six Negroes with food and drink started for Makerare. A wide brook, which has its source up there with water between 13 and 15 degrees, was the goal, under which they were to place our beer and other beverages along with the butter. After we drank our tea and ate what we could, we mounted our donkeys, and, with me in the lead to show the way, we began our excursion.

Between here and Makerare there is a good piece of genuine Africa, that one seldom sees so close to the coast. Completely uninhabited by people, wild animals and birds had taken up residence here. What singing, chirping, peeping and screaming of birds is not heard here. At the break of dawn one soon hears the cluck of the cuckoo, soon the wood pigeon's cooing, soon the rhinocerous bird's scream and soon the song of one of the small song birds, which is not as beautiful as that of the thrush. But here, where it sings for grateful listeners, we think it beautiful. There is then a humming over our heads, and soon there comes a flock of small, green parrots. They are very shy and were disturbed by our talking and laughter. Now the small, black birds with red breasts and heads fly by. Then came a blue, yellow, red, white mix. Next there was a little bird, only two inches long with only two feathers in the tail six inches long. It is an irritable little character, who does not tolerate other birds near him and is always occu-

[10] Kempinski was one of the most famous restaurants in Berlin around 1900. Regarding Mrs. Nimb's cookbook: see Chapter 6, note 12.

No text and no date, but obviously an outing

pied with chasing those who come too close. I must not forget the little hummingbird, which one sees fluttering around every flower. High up in the air one sees the king of birds, resting on his mighty wings, float through the air with his eyes towards the earth, looking for food. "But what is that? Are they people." This from a brand new African. In front of us was a flock of dog monkeys. They have this name, because they express their feelings just like a dog. The monkeys did not seem to care for us at all, but ran slowly and played with each other about 30 steps from my donkey. They are very large animals, and measure about five feet, when they stand on their hind legs. A spotted dog, who was with us, noticed the apes and went full speed in among them. It came close to being a costly experience, because a couple of male apes took hold of it. If we had not rode in and helped it, the dog would have been bitten to death. The same apes are a horror for the Negroes, because they steal his maize and *muhogo* in the fields and kill his children if they come too close. We then saw a number of antelopes and the tracks of three elephants, who have their territory here. At about nine o'clock we reached our goal, Makesane Brook. I had been up there a couple of days before and made a little shelter about 16 metres square. Here we spread out our two mats and laid down in the shade. How good the beer tasted that day!....[letter ends here]

[it seems CL must have returned home during 1900, and was in Berlin on his way back to Africa]

Berlin, 28 November 1900

Dear Vilhelm [Carlsen],
I bought the electric torch yesterday, which I hope you will receive in a couple of days. When the light is not too often used, the apparatus can work for two years. A new apparatus is set into the torch by pushing down the ring that sits on the torch and removing the two screws that hold the handle. The handle is removed, and the new apparatus is set into the torch, which is hollow. As far as the thermophor is concerned, I have not yet been able to find where it can be bought or what it is. The other night, at 12, I was in an automat restaurant, and there I saw written "thermophor hot dish" But, when I asked what a thermophor was, the people there did not know. But now I will find out. Up until now I have had so much to do. Even though I have gone down *Friederichsstrasse* from one end to the other, I have not found any thermophor factory. I will be traveling to Naples in a couple of days and will be sure to send you the description and treatment of sisal agave.

It is too cold here, and I have no peace and quiet.
Warm greetings to everyone
Yours affectionately
Chr Lautherborn

Onboard the steamship *Gouverneur* one day from "the Line" [equator]
December 1900

Dear everyone!
Allow me to first acquaint you with my travelling companions and the ship's officers. The captain, naturally, comes first. He is 31 years old and has a stomach so big that it always comes through a door two seconds before him. The easy life he leads is the reason he looks like he does. He is, like all fat people, very good-natured and always has a smile on his lips, when he wishes us good morning. We then have three officers, who all have so much to do that we seldom see them. At my table there sits a mission director, Mr. Schwartz and his wife, a couple in their 60s. They are now travelling to Africa to see the various mission stations. There is a railway director, N., who is going out there to make an estimate for the new railway, which has now been approved. He had been in Cameroon for six months and is terribly clever, but he says very little. We are only four at our table; i.e. counting my insignificant presence, you have them all. At the next table reigns the

fat captain at the end of the table, then there is the Dutchman, Count S., who was with the Duke of Abruzzia at the North Pole. It was a pleasure to hear him talk about his polar bear hunt, when we were on the Red Sea and had 36 degrees heat under a double awning; one could feel a chill go down one's back. There are then two doctors, a captain, two merchants, a salvage director, and that table is finished. The next is international. Here there are Portuguese, Italians, Englishmen and Dutchmen. They are all people that I have nothing in common with. I would only exclude a Dutchman, Mr. M, with two daughters. Especially the last are very interesting. The table behind us is all English. One of them, Doc S., shares a cabin with me, and we are very good friends. Another table consists of only Portuguese. Two of the tables are not used, since not all of the cabins are occupied. We had a couple of terrible days on the Red Sea. It was so hot that the pitch, which is pressed down between the deck planks, melted where the sun shone on it. As a result water had to be thrown on the deck constantly to prevent the pitch from running off. It has been very quiet. Not one evening, while we were on the Red Sea, was there a dance or a concert. But, within the last few days, we have made up for it. It is now quite cool, and every evening first and second class assembles for a dance; and it goes very well from nine to eleven. The two little Dutch girls dance very well — and never become tired, so I am sometimes more waltzed than I like. Well, motion is healthy, and it has not hurt me any.

I made the acquaintance of a Miss Lola on such an occasion. Since she had an excellent voice, she was invited by us to come over to first class in the evening, when we did not dance, and sing for us. She was glad to do it, and I have to say that she sang quite well. However, it also occurred to me, when I observed her, that it was possible that she had had a bad past. Furthermore, from the way the various gentlemen talked to her, I had to assume that she was *eine Dame der Halbwelt* (Demi-monde). At the dance the next evening I acted as if I did not see her and danced with the two, little Dutch girls und *liess Sie lindes liegen* [and let her alone]. But now comes the strange part. When the dance had stopped, she asked me for a ten minute conversation. I agreed, and what she told me was almost a novel, which, since I have the time, I will tell you too.

She began: "I was born in Vienna. My father is a quite wealthy man. I completed the gymnasium and learned to play and sing. At a dance I met a young, handsome man. We danced a lot together, and he, through others, became acquainted with my father. We saw each other often and developed a burning love for each other. He proposed to me, and I told him that he should speak to my father (her mother was dead). He did this but received a denial, because my father did not know enough about him; and because I was too young. But he waited to see me every time I took my walks. My father had forbidden him to come to our house before he could clarify who he was and where he came from. He filled my ears with sweet, flattering speeches, kissed my tears away from my eyes when I, cryingly, asked him to go away. I did not dare have anything to do with him because of my father. "Yes, Rosa," he said, "your sense of duty is much stronger than your love for me. If that was not the case, you would not think about your father's prohibition. The joy of being together would drive all such thoughts away." I was young then, only sixteen. I am now twenty, but the last four years have brought me sorrow, shame and misfortune. You cannot see this. You see by my dark hair and my brown eyes that I have a great deal of the hot, Italian blood in me. Since you have been in Italy, you must know that the people there are made to sing love songs, to speak sweet, flattering words, to forget everything in the world for an evening together with one's beloved; and that was my nature. My mother, a warm-blooded Neapolitan, married father during one of his trips to Italy. She ran away with father, because her guardian would not give his permission for their marriage. Mother's nature was in my blood. My beloved, who continued to pursue me, talked me into going with him to London. "We can be married there." he said, "and, when we return, your father cannot separate us any more." What could I do? On one side the strong sense of duty I had to my father, on the other side a young, handsome man who, in every movement, every word, every sigh, talked only of love. My warm nature could not resist him. I agreed, and we went to London. We were married. I wrote a letter to my father telling him everything, begged for his forgiveness, and asked if he would write to me at a Paris hotel, whose name I gave him.

When I arrived there, there was a letter from my father. He asked me to never darken his door again and disinherited me. My husband was very upset by this, but he said that we had each other; and nothing on earth would separate us. We went to Johannesburg in Transvaal, and I soon discovered that my husband was a poker player. He tried to make me believe that he was employed with the railway and had night duty, but it was not long before I knew what he did at night. I tried to make him stop. I said that it was wrong that he had lured me from a good home out into the world, when he could not, in an honest manner, support me. I said that I would rather take my life that day. He said, "I have never worked. I have only earned my living by playing cards, but, for your sake, I will try to earn my living as an honest man." When he had promised me that, I would not talk about it any more but let him try to get himself together. So it went for two years. We had a good house. We lived and dressed well. We were satisfied, but I could not see any sign that he had changed his life. However, I remained silent, always hoping that he would honour the promise he had given me.

One evening he brought a couple of friends home. I played and sang for them, but was surprised at the way they treated me. After they had left, I asked my husband not to bring such friends home again. I did not like the provocative, indecent looks they gave me. If he still loved me, he should spare me from the presence of such people. "But what do you think", and here she began to cry, "my husband answered?" He said, "What good can it do that I love you, when we have nothing left to eat; so will all love soon disappear. It has gone badly with me. I have lived on credit for the last half year. My gambling has gone badly. I have not been able to get other work; and we use money to live. Why cannot you help me to earn money? You see, these gentlemen I brought home are very rich. You know that in Johannesburg there are almost no white women. If you will do what they want, you can earn huge amounts. Our poverty will be gone with one blow, and we can live in luxury." Here she took a little pause, tried to stop her tears, and continued again. "You can imagine how this speech affected me. This was the man to whom I had given my love, who now advised me to sell my body to the first who came along. The thought was unbearable.

I do not know what I said to him. I only know that I went out that night with the resolve to drown myself. But, when it came to it, I did not have the courage. Another thing was that I carried under my heart a three month old baby. I did not think it was so bad to kill myself, but why should the creature be killed who had not yet done anything wrong. I went home again. I asked and begged my husband to look for a position. I would do everything to help him. It was only what he wanted me to do that I would definitely refuse. But my husband now showed a completely different side of himself. He said, "Do you think that I would take care of our support alone. If I had had luck in gambling, I would have been glad to. But, now that luck is no longer with me, you will have to help me. I only ask that you be reasonable and listen to what I say. You want me to seek work. What could I earn with that? I never learned anything in my life. My father was a poker player, and I have also become one. I would have to work as an ordinary day labourer to earn a living, you see. I can and will not do that. You say you will try to earn money in an honourable way. Go ahead and try and you will soon see how long you last. As a prostitute you can earn more in one night than in two months as a worker, perhaps as much as in one year. I know the rich men in this town. I will bring them to you, and you only need to demand money; and you will get it. One hundred pounds sterling means nothing to these people, when you are willing to please them."

You can believe that I tried to resist my husband's will. I went out the next day to seek work, but I found nothing suitable. I continued to look for three weeks without result. My husband saw to it that we had company every evening. After spending a couple of hours in our house, they went away and left me alone with one man or another. He would always try to catch me in his embrace, but I firmly rejected the advances. It had come so far that we hardly had enough to eat. I had not gotten any clothes in the last one and a half years. When I asked my husband for money, he yelled and cursed. "Earn it yourself; you will never be able to earn it easier." The constant pressure from my husband, our poverty, and to earn money for when I gave birth made me, one evening, in my despair, give myself to a rich

Englishman my husband had left with me. He paid a huge sum.

It went with me as it does with so many others. When the first step on the road to vice is taken, the next step quickly follows. After I had earned an amount large enough for us to live on for an entire year, my husband said that I now would go to Zanzibar to give birth and thereafter go home and live a decent life. We went to Zanzibar, but my husband, with nearly all the money, left me. I stood completely empty-handed in Zanzibar, due to give birth in a month. I got a job in 'America Bar.' I played piano and sang and earned fifteen pounds per month. I gave birth in the French hospital. After I recovered, I continued with my piano playing. I got a Negro to take care of my child in the evening. Naturally many men came in who wanted me, but I always said no. I earned enough for myself and my child by playing the piano, and I took walks during the day. My child is now 1½ years old. I have been in Alexandria, where I was divorced from my husband.

I am going to Tanga to be married with Mr. S. You know him. He is an official for the Italian company out there. He will be a good husband for me. The authorities in Alexandria wanted to give my husband two years in prison, but I asked them to let him go. I only wanted to be divorced from him. What he did to me no authority on earth can make good again. Oh, how low I was! Thinking about it has nearly driven me insane. But I will, from now on, lead a decent life, and nobody will be able to say anything about me. I will also be able to look into my child's eyes calmly , when it grows up. The young man I met in East Africa has always been so good to me and has always helped me. I did not want to go back there. I wanted to stay in Alexandria, and there try to support myself and my child, but I was given no peace. Therefore I am going back to marry him. I know that he is honest, and faithful, and a good worker, and that he will take care of me, you see! I told him that I could never marry, because any decent man would want to know where my father lived, what he is etc. Since my father's house is closed to me, there is no one who will have anything to do with me. Only Mr. S, from the first moment, has loved me. Even though he knows how I have been, he would not give me up. Naturally you wonder why I am telling you all of this. But I am alone in the world and must confide in someone I can trust. There is something so open and honest in your face that I thought I would ask you. Am I committing a sin in marrying the young man. Will it not prevent his advancement? Answer me honestly now that you know my past." "It is a singular matter to judge such a case"; I told her. "If your husband is a man of a good family, I believe that you will make him lose his family just as you have lost yours. That is to say that his parents will not know anything about the villain of a husband you have now freed yourself from. They risk that they will be treated by him in a similar manner. The only right thing for you to do is to say to Mr. S that, for his sake, you will not marry him. It is enough that you are ruined in soul and body. He should not also be pulled down. If you love him as you say, so tell him, when you come to Tanga, and take the next steamship to Alexandria. Since you have been in Zanzibar, your life is, of course, known on the coast and your young man risks losing his position, as soon as it becomes known that he has married you.

"If you and Mr. S. can find a position in a part of the world where you are not known, so marry him I say. But do not do it in Tanga. It will be his ruin." She became very pale. The paleness was made clearer by the blue-black hair and eyebrows. The tears flowed down her cheeks. But she put out her hand and said, "My deepest thanks, your advice shall be followed, no matter how difficult it will be for me. I would also ask you not to think too badly of me. Oh, how much I would have liked it to be different."

You can imagine how hard it was for me to tell the poor woman what I honestly felt. But I know people in Africa. It would not have been long before Mr. S. lost his position. She could have had as good resolutions as she wanted; everyone would still have said, "she was one"!.

Well! That was a rather long story. But I thought you should hear it in your quiet, little corner, where, thank God, there are no dark clouds in the sky, where the sun is bright and clear. One first feels how good one has it, when one hears about others' misfortunes. Give my greetings to mother. I hope that she gets well and heartfelt greetings to all of you.

from your
Chr Lautherborn

CHAPTER NINE

KIKOGWE AND SISAL: LAST YEARS IN AFRICA

German East Africa Company, Kikogwe
26th January 1901

Dear Sister
Thank you for your letter with the Christmas greetings. It was the first that I received here. It is true that Axel had promised to write me at Christmas, but he probably has other interests. I arrived here, as I said, on Christmas Eve, and it was held in Tanga with a really magnificent Christmas tree, which we had with us onboard. The heat was tolerable, but I thought a lot about you with your beautiful, white Christmas. It is not the same here.

I am glad to get back to my work. Now it is up at five, if one wants to keep up. But that is important for me. Without work life has no interest. There was always something I missed when I was home. It was not having anything to be responsible for. But you can believe that I have that here. At this time my assistant, Meyer, is in Kikogwe, sick with pernicious fever, and I drive down here every morning to see to the machines, because we make hemp every day. The machines make such a noise that I am quite sick in my head from it. But it goes away, when one becomes used to it. My only goal now is that my plantation shall be the first to show a profit, so that I can stand as number one. That this will be the case is fairly certain, if God continues to give me strength and health which I hope He will. I never talk here about how I believe the result will be with the plantation. The reason is that it has cost me too many years of hard work finding out what will grow here to give my experience to all the people, who just sit here and wait for a good idea to grab with both hands. When they have gotten it, they claim it as their own and try to push me down in one way or another.

I take the opportunity every so often to run away from the machines and write to you. I do not, however, get much peace, because, at the speed we are going, there always must be a person standing by. I would now ask that, when you write, you address the letter to Pangani German East Africa not Mwera, Zanzibar.[1] I always get the letters three weeks later, and it is con-

[1] The Deutsche Ostafrika Linie was highly subsidized by the Imperial government and provided regular fortnightly service between Africa and Europe from 1890. Later bigger steamers sailed every three weeks, and an intermediate line came every six weeks. The post was carried on these liners. H. Brode, *British and German East Africa: Their Economic and Commercial relations*, London 1911.

No text and no date. CL with two of his assistants and a locomobile

nected with great inconvenience. Is Kirsten in some shop or other in Hjørring? She writes so briefly and business-like. You can believe that the big umbrella I bought in Hjørring has been of service to me here. When I came to Tanga, and the Christmas holidays were over, I went by *dhow* to Pangani. When I asked the captain how long it would take to get to Pangani, he said four hours. Therefore I only took a couple of bananas and a couple of bottles of mineral water, but we took 48 hours to get here. When I arrived in Pangani, I was so hungry that I could hardly walk. I spent the two nights on deck. I slept on an old sail, which had been folded for me. But at night it began to rain. I opened up my big umbrella and set it fast in the edge of the ship, so that it covered my upper body and my little [unreadable] to cover my legs; and I slept without getting a drop of water on me. Now I use it every day on my carriage, and it protects me from the sun's rays as well as from the rain.

When I go to Zanzibar, I will try to see if I can get the plates for you. Now your Christmas is also over with, and, no doubt, you have gotten much happiness out of it, especially the children. For them it is, of course, a special celebration. Its religious origin, for them, is a minor thing, as long as the Christmas cakes and pastries are good, and the Christmas goose is not forgotten. Well, now I have been here eight times and written a little, but now I have to stop. Do not forget now that my address is Pangani Deutsch

'My donkey cart'

'My house in Mwera seen from the side'

Ostafrika, so I will get everything quickly. The most heartfelt greetings to you and your children with wishes for a good and happy New Year. Give my regards to Vilhelm [Carlsen].

Your affectionate brother
Chr Lautherborn

Deutsch-Ostafrikanische Gesellschaft, Mwera
17th April 1901

Dear Sister!
Your letter of the 4th of March reached me here a couple of days ago. I was in the field, when I got it. It is now the big rainy season, and the rain pours down as if it were the Biblical flood. That is the reason that I have begun to write to you today. It is pouring down in buckets, and, if I cast my eyes down on the field from the hill I live on, it looks like a large lake. The rain is falling so strongly, that, even though the field is well-drained, it cannot handle the water. Yes! What do we actually know about rain as it rains here? At home it is nothing new that it pours down in buckets. The only difference is that the buckets are 4 to 5 times bigger here than at home. I have seen it rain for six days and nights in a row. When the rain has finally stopped, the water has cut deep ditches in the earth; and a flat field has suddenly become a mountain and valley landscape in miniature. But one learns as long as one lives, and I have also learned to drain the water out of the field without changing its form. When I got your letter, we had just had the first, real rainy day this year. I was very pleased over it, because we had just planted 20,000 sisal plants,[2] and they had become quite dry and withered. But, when the rain came, they immediately livened up. You know that I bought a large umbrella on Springvandspladsen in Hjørring. It is now my close companion every day. On a rainy day I would not give it away for its weight in gold. You can believe it was a great success with the natives. They have offered to work for me for three months, if I would get them such an umbrella.

They believe that, when they travel from the coast to their home, they will not need to build huts to sleep in at night. The only thing they will need to do will be to open up the umbrella and set the shaft in the ground, and the hut will be finished.

Today is the 26th, so you can see that the letter has lain for nine days, without me having time to write more on it. The rain stopped, so I had to

[2] Sisal was first grown in GEA at Kikogwe in 1893. Lautherborn experimented with sisal and managed to prove it was a viable commercial enterprise. By 1902 sisal was the third commercial crop in value in the country after coffee and coconut (Koponen 1994:206). By 1913 sisal was the biggest export for GEA (Sunseri 2002:19).

go quickly into the field to get the water out of it. Vilhelm would like to hear something from me for the newspaper, so I will now try to get time to write some day or other. But I think that, when he has such an interesting person as the Harboøre priest[3] to write about, it should be enough. Yes! I think it is high time that he came. Otherwise half of Denmark would have gone to hell with Home Mission. He is my man. They are ideas I have had about religion, since I was old enough to think for myself. I hope he will accomplish something with his speaking. I would only wish that, when I visit Denmark again, the cloud of anxious expressions from Home Mission supporters, who only fear the Devil and hell, will have cleared up; and that they, with a calm and happy expression, can look up to the God the Harboøre priest preaches about. That is, in my opinion, the correct one.

Just think, I got a letter from Cathinka. The age of miracles is not over yet. How else could her husband give her 900 kr per year after he is dead. He must really love her. Otherwise I cannot understand it. It is just such weak men that easily fall in love with a big woman. But thank God that she is well. How is it with Kirsten? Is she to be a Cinderella and stay home to sweep and clean, while Ellen goes to a folk high school?[4]

I think you have used so much money on her in Randers to learn housekeeping that you do not need to spend money again on her in folk high school. Furthermore why send her to la Cour?[5] It is probably because it is so far away that Denmark ends. She could not go further away, if she wanted to use the Danish language. Besides I do not understand why you send your daughters out to learn housekeeping. After all you have such a large home that you could very well teach them at home. As far as mental education is concerned, I think that one could seek far, before one found a pair of parents, in which the mental and spiritual education is on such a high level

---

[3] Harboøre was a stronghold for the Home Mission in Denmark. The text must refer to Anton M. Jensen, minister in Harboøre 1899–1901. Before him the parish for several years had had ministers that belonged to the Home Mission movement and in January 1901 Harboøre parish once more got a minister that belonged to the Home Mission. Anton M. Jensen, however, did not belong to the Home Mission and had collected money and books for a library, as he wanted to educate the people of the parish (in a Grundtvigian sense of education). Late in the year 1900 lightning struck and burned down the vicarage, including the new library. A newspaper related to Home Mission wrote that the lightning perhaps was a sign from God? (*Dansk Præste- og Sognehistorie VI*, Viborg Stift, Kbh 1963 p. 614–615).This provoked a public debate in Danish newspapers about the situation in Harboøre, the Home Mission etc. Vilhelm Carlsen and *Vendsyssel Tidende* seems to have taken part in this debate and, like CL, must have sided with Anton M. Jensen.

[4] Ellen Carlsen born 1881 and Kirsten Carlsen born 1883, daughters of Caroline and Vilhelm Carlsen. Cathinka was CL's youngest sister.

[5] The Folk High Schools were aimed at adults, and a vital part of the Grundtvigian movement. The Folk High School in question here is Askov Folk High School. Askov belonged to the folk high school's 'ivy league' and was for a number of years the most prominent among them.

as with you. If only Vilhelm's madness could have a little of your calm nature and you a little of his passion, you would be a pair of excellent educators. I know that you learn from and share with each other. Could not you also share tempers? Then there would be a little more passion on one side and a little less on the other side. Yes! Vilhelm believes that his family has stepped into danger by living outside Hjørring. I remember well when we measured the land near Lønstrupvej. At that time he told me so many horror stories that the hair would have raised on the heads of most, also on mine, if I didn't know that his fantasy helped him well on the way. Therefore he has probably, by now, acquired a dog so large that his growling can be heard in Hjørring. But he will not likely have use for it. It is such a quiet and good neighbourhood to live in, that criminals would never go there.

I live here in the jungle. Up to my veranda there are wide steps, but there is no door for them. My door out to the veranda is always open, so that I will have fresh air in my bedroom. By my door lies a little, black spitz, who makes a terrible racket every time someone comes up the steps. Unfortunately I have gotten so used to its growling, that I no longer wake up. But it could never occur to me that someone would hurt me. But that is out here in an uncivilised country. At home civilisation is, of course, on a higher level. One cannot say, however, that it is, therefore, safer than here.

Wish Hjalmar and Marie well with their little son.[6] I have darn little time to write to them. Yes! Now it might have been better, if I had written a newspaper article. But I do not have any time right now. But I, for sure, will get something finished. There is enough to write about here, if one just had the time and the desire.

Warm greetings to all Your affectionate brother
Chr Lautherborn
7th May at 11 am
The letter goes out tomorrow.

Mwera, 18th November 1901

Dear Caroline
Even though I was really glad to receive your last letter, I cannot say that I was happy to read it. What you wrote about father was extremely sad—if such a strong and big man should die of something so insignificant. I have now inquired about the problem Father has, and have learned that it is not so bad, if one will just submit to surgery; but this he, unfortunately, will not do. It is, after all, nothing in our day. Surgery is so developed, that an operation would be the same as an uneasy sleep with vomiting when

[6] Hjalmar Carlsen born 1875 was Caroline and Vilhelm Carlsen's oldest son.

one awakens. I think one could submit to this, when one could have one's life lengthened by 20 years. But perhaps he doesn't know himself how bad it is. I think he should be told. It would be difficult to do by letter. But cannot you get him into the hospital and have him operated on, if it would do him any good that is? If I was home I would do what ever I could for him, but, unfortunately, I am here. I will write to him today, when I am finished with this letter, also to get Axel to America, because it certainly is about time. You must not think that I have not written to the old folks about it, but I then got a letter from Mother saying that she could not let Axel leave now.[7] She was going to tell me why, but I have yet to hear what the reason is. But now I will see if I cannot get him to go in the spring, because, at this time, he will have worn himself out too much.

Dear Caroline you write about your many children and do not know if you are able to give all of them a good upbringing. But just let me give you some good advice. I have seen, during my visits to Denmark, that you work far too much yourself and give your bigger children too much freedom. You have, after all, both Ellen and Kirsten. One can only say about them both 'a couple of nice and pretty girls', but why not use them in the house? What would be more reasonable than that one took care of housekeeping and the other, together with you, the raising of the children? Instead of wearing yourself out, you should sit nicely dressed and with curled hair, only, now and again, like a commanding general, have your troops pass in review and correct any, possible errors.

I now have nine hundred Black children. In any case, they all call me Father. But, if I tried to take care of each one, as you do, I would not be able to do anything. No, the older ones, who have been with me for a long time, must take care of the younger. In this way I train them for work. My dear, little sister that life is not what it has been for you must not be ascribed to Astrid's death.[8] I, for example, have not had the sorrow you have had, but life still does not have the same value it had in our youth. We have both reached the top of the hill and have begun, with small steps, to go down. We might still be able to see over the top to the other side, where we have climbed up. It is, however, so unclear to us, that we would rather turn our gaze towards the side we are on. But you, with all the youth in the house, should actually not get old. I do not understand what you say about the Christian faith. I always thought you were as believing as you wanted to be. Every intelligent person believes in God. So do you, even though you do not go to church every Sunday. What more would you do? I cannot believe that you have begun to look at Home Mission.

I know that you live in a good place, and I can imagine that there is healthy

[7] Axel (also spelled Aksel by CL) was the oldest son of CL's youngest sister Cathinka.

[8] Astrid Carlsen born 1879, daughter of Caroline and Vilhelm Carlsen, died in childbirth 1900, giving birth to her first child.

air for all your children. It must look beautiful, when the white snow lies on every branch and leaf in Svanelunden [park in Hjørring], and the sun stands up and casts its beams over it all. That is how I remember winter in Denmark and Christmas, which was always white. What a wonderful time that was with Christmas goose, pastries and cakes. It is also actually the time when I feel homesick, when Christmas comes. But it will be many years before I spend a Christmas in Denmark. [A short passage discussing what would be suitable Christmas presents for some of Caroline and Vilhelm's children has been left out.] And so I wish you all a really Merry Christmas and a Happy New Year.

Your affectionate brother
Chr Lautherborn
Do not forget Karlsvognen [Danish name for the constellation the Great Bear—but here it is probably meant as a nickname for Vilhelm Carlsen, Caroline's husband]

[CL is starting what will be his last trip home to Denmark]

Pagnon's Hotels on the Nile
The Luxor Hotel, Luxor
28th April 1903

Dear Caroline!
I have now been here in Egypt for about fourteen days. What I have seen in this time surpasses everything. One lives here 4000 to 6000 years back in time, when heathenism stood at its highest. So much of what I heard in school but did not learn comes back now in memory. I do not think that even you could pass an exam about ancient Egypt. Even though I, as you know, am no scientist, this trip has been incredibly interesting for me. I am not thinking of holding a lecture about the ancient, Egyptian gods, but only to tell you a little about life here. Let me begin with the worst annoyance, dust. I have travelled more than a little, but such dust as there is here I could not have imagined in my wildest fantasy. On the trip from Cairo to Luxor the dust lay half a tomme [inch] thick in the coupe. I was so covered with dust that I resembled a wrapped mummy. I could not get air through my nose. I could not see, because of dust grains in both eyes. When I took in air through my mouth, I could blow out a cloud of dust like people who smoke cigarettes and exhale the smoke. I arrived here in Luxor at twelve midnight, after riding in the dust from eight in the morning. When I had bathed and gotten into bed, I could not get my breath. After throwing up about a bucket of desert dust, I could finally find rest in the arms of Morpheus. What I have written is not an exaggeration but the pure, unadulterated truth.

The next annoyance is the flies. Where in the world do they come from?

Touristing in Egypt. CL is riding the donkey

Here every person is equipped with an instrument to chase flies away from their faces. It is a baton about two fod [feet] long. In one end the long hair of a horse's tail is set. With this every man on the street is seen fencing to keep the flies away. The eyes and noses of the small children are always full of flies, and they do not seem annoyed by them much. That is probably the cause of the widespread Egyptian eye disease. Small children! No place on earth have I seen so many as here. The reason is probably the healthy, dry climate. Where else is it as dry as here! A little while ago I bought a box of cigars, and, when I bit the end off of one, it just crumbled into dust in my mouth. When I asked the cigar dealer if he had not found them in Ramses II's grave, he assured me that he had not had them longer than three months, which I have to believe, even though it is hard. The heat here is terrible. East Africa is nothing in comparison. I will, therefore, hurry to Cairo, where, at least at night, it is bearable. You can tell Mother that I eat strawberries with cream every day and think about her each time. If they could take the transport I would like to send her some, but they cannot. Now dear sister you must excuse me. My three, travelling companions are complaining because I write so long, and we shall again go back 6000 years in time.

Live well and give my greetings to all from your affectionate brother
Chr. Lautherborn

Berlin, 17th May 1903

Dear Caroline!
I am now in Berlin. Have here a great deal to do with the company. Thinking about leaving here on Wednesday morning. Will go to Randers and stay the night there, and so will be home in Denmark and with you in Hjørring on Friday or Saturday. I cannot say that I enjoy being here in Berlin, but I have to wait upon Director Lucas. Otherwise I will have to come down here, when I have come home, in order to get my new contract.

Tomorrow morning, Sunday, I will go to Stettin and visit Emil to get away from all the noise here. But have to be back on Monday morning, when I, with Director Winter, have to buy machines for East Africa. I long for peace and quiet and to be able to relax a little. When one has travelled nearly every day for six weeks, one is tired of travelling. Give Mother my greetings a thousand times. There was a letter waiting for me here from Cathinka. I was thinking about going over to Copenhagen but will not do it now, since my stay here has been longer than expected, and so it will be that much longer before I get any peace. In Randers I will try to meet Buchardt, if I stay there long enough. In any case it will only be for one night, in order to sleep well before I come to Hjørring. Otherwise the first day will be sleepy and yawning.

Greetings to your children and your husband and greetings to you from your affectionate brother
Chr Lautherborn

Berlin, 7th September 1903

[On his way home to East Africa again]

Dear everyone!
Arrived in Berlin yesterday at four. I could have been here at twelve. Was too hungry when I arrived in Hamburg and was there for four hours. Berlin looked abandoned and empty when I arrived, perhaps because everyone had left. At eleven in the evening, however, when the theatres were finished, one could hardly get through the streets.

When I arrived this evening, it was 36 degrees Celsius, and the same today. I dreamed last night that Vesuvius had erupted, and the glowing lava fell down on me. I woke up and cast off the blanket and slept very well until eleven in the morning. After sleeping two hours until supper I feel like I have made up for the night between the 5th and the 6th, when I slept badly. Yes, what a difference there is between Hjørring and Berlin. Here one nearly has to have an eye on every finger, because buses, carriages, motor cars, work wagons and bicycles whiz by each other. At each street corner one crosses one has to look carefully, otherwise one will be driven over. This happened to a soldier near Friederichstrasse's train station tonight. He was lucky, but he got a few bad scrapes. I stood in the passageway between Unter den Linden and Behrenstrasse today and read on a sign that one could have one's eyes scientifically examined, if one would buy a pair of glasses. Suddenly a girl ran into my arm. "*Kommst mit Maüschen* [come along little mouse], she said. But I answered, "*nein am Tage sind nicht alle Kätchen grau*" [No, at day not all cats are grey]. Since she did not turn around, I could not see if she was happy; but I do not think so.

Tomorrow I will go to the company. I am practicing speaking German, because yesterday it did not go so well. I surprised the man next to me in the theatre by always addressing him in Danish, so that he, in the end, began to address me in English, but it is going better today. One does not see the 'Kaiser beard' very much here now. Most men remove it completely. Ladies' hats are as flat as pancakes and quite large. The dresses are 'Mother Hubbard' cut.[9] The shoes are, in this heat, red or yellow. If I had had my tropical clothes here, I would have worn them, but, unfortunately, they are in Naples. My address is Berliner Hof Berlin W. I did not get private lodgings, because my driver was drunk. Since I live here for four marks per day, I am, I believe, much safer here from thieves than in lodgings and will stay. Tomorrow, when I have been with the company, I will go and buy a carriage for Mother and also get hold of a [name of a company unreadable. Most probably a sales] catalogue for Otto. Today I am just taking it easy. I will go to the theatre tonight. I have now found out where one can get non-alcoholic drinks, and, when I come back from Africa, I hope that Vilhelm and Caroline will meet me here.

Best wishes to all
Yours affectionately
Chr Lautherborn

Berlin, 18th September 1903

Dear Everyone!
Before I leave Berlin, which, I can assure you, will not be in the best humour, I am, however, somewhat calmer, since I know that there is a little improve-

[9] Mother Hubbard is a figure in an English nursery rhyme. A 'Mother Hubbard dress' is a frock with a wide cut held together around the waist with an apron. It has long sleeves and a high neckline, so it is altogether a dress that definitely does not set off the body.

ment in Mother's illness. I cannot say, however, that I have the best expectations. It is, however, a comfort to me that, if she should die, I went to Hjørring, pressed her hand and kissed her for perhaps the last time. So do not let anything bad happen to the unhappy lout and telegraph manager. I believe everybody feels calmer, if he has done his duty, and I feel good for having been there. If, before the 23rd, there should be changes in Mother's illness for the worse, I ask you to telegraph to Lautherborn, Müller's Hotel, Naples. If, on the other hand, she should begin to get better, then don't. I ask you to give her my most heartfelt wishes. It is terrible that I have to leave now, when she is sick, but it cannot be otherwise. My deeply felt thanks to all of you, who care for her so well. I am leaving here tomorrow and do not believe I will go to Stettin this time.

The most heartfelt greetings
From your
Chr Lautherborn

Mwera, 22nd March 1904

Dear Sister!
Your letter with the announcement of Mother's death found me sick in bed in Tanga. When I received the letter, I already knew. *Vendsyssel Tidende* had delivered the news. Sad I should say, but I cannot, because I know Mother must have been a burden for herself and those around her. Death must have been a blessing for her, not to mention all of you. It must have been especially hard for you dear sister to take care of her. I can well understand that you feel relieved that she died so quickly and now has peace. I have, since I returned to Africa, read the obituaries in *Vendsyssel Tidende*. I did not believe Mother would live long, and, strangely enough, I received the announcement through the newspaper, before your letter reached me. I am, however, glad that I came to Hjørring that time. I managed to say farewell to her then and know that I have done what I could.

It is first now that Lilly tells me that she is married, and I also know that Cathinka has been sick. It appears that the barrister does not live up to the match Lilly has made. Nothing has been written to me about it, but I have read between the lines. What is that poor girl doing with that man?

I do not believe in him. He may be alright, but, in my opinion, he is too fat and has too little desire to work. Ask Otto to send Lilly 200 Kr from me, so that she can buy what she wants. Let me now ask you to take life a little easier than you have up to now. There is no use in getting old before one's time. You are strong and well now. Try to keep up your strength. After all you have young children. Let them take most of the work from you and have a good rest. Work in the garden. It is healthy, and you will see, you will become young again.

It is going very well for me again. I was sick, but it is over with, and I am very occupied with setting up my steam engine. I am now also a ship owner. I have a two-masted schooner, which transports hemp from Pangani to Tanga. I will, when I can, send you a picture of it under full sail. Here on the plantation everything is going better than I expected. This year I will send 1,600,000 pounds of hemp to Europe. I will then have 150,000 Mk profit.[10] Next year it will be approximately double. It really provides good motivation to work.

Greetings to all acquaintances in Denmark and also to you from your affectionate brother
Chr Lautherborn
Did you get the silk? I will send more.

Mwera, 27th August 1904

Dear Sister!
Thank you for your letter, which reached me a couple of days ago. My birthday was already a little in the past, but that doesn't mean anything. So the box with the antelope skulls has arrived. But it was not the intention that they should sit up in the attic. For hunting enthusiasts they are rare, and I promised them to Mr. Lønborg Friis.[11] Ask him if he would like to show them at the museum until I, someday, come home and want them myself. He can also have the mats. If you would like some, so take them and send him the rest. There are certainly many beautiful and rare horns, and I believe they will draw visitors to the museum. So the silks also arrived. But tell me if they are good. I have not seen to it myself. Are they better than the silk one buys at home?

Yes! Who could have been in Lønstrup [fishing village on the North Sea, close to Hjørring]? That would have been something for me, but I have to wait two more years yet. While I am sitting and writing, I can look over the Indian Ocean over to Zanzibar. But the sea is so quiet and calm and the coast so green and lush, that I can hardly compare it with the uneasy North Sea and Lønstrup's naked cliffs. It is Saturday today, and the market place near the workers' quarters are filled with men, women, young girls and boys, all selling foodstuffs. How often I think about your children, when I walk

---

[10] 1000 Mark equals 49,054.70 D.kr. today (2004) The 150,000 Mark CL thinks he will earn on hemp this year (1904) equals what today would be 7,358,250 D.kr.

The 250,000 Mark mentioned in his next letter equals 12,263,675 D.kr. and the 600,000 Mark equals what today would be 29,432,520 D.kr.

[11] Mr. Lønborg Friis (1852– 1913). Dentist, amateur archeologist, one of the founders of Vendsyssel Historiske Museum, served as daily manager (unsalaried) of the museum 1900—1912 (see also chapter one).

A picture of the Carlsen-family taken in Lønstrup when CL was on 'home-leave', summer 1900. Vilhelm Carlsen stands in the back row (without hat). Caroline is sitting in the second row, number two from the right (with a small black hat). CL sits in the second row on the left side. The rest of the flock is the Carlsen children—the oldest of them together with their spouses

across the square. Here there are fruits by the thousands. I would love to have them here and pour as much fruit down them as they could hold. If the fruit could only last, I would like to send them home; but they would rot on the way. Here there are oranges, pineapples, bananas, stink fruit, a variety of African pears, *Ustaferi* (the African strawberry), gooseberries and blue, African plums (*sinbaran*), elephant lice, also called *malibues*, a beautiful fruit, which has something in common with our Danish apple. The nut sits outside the fruit and has a great resemblance to a giant louse, therefore the name elephant lice.

Saturday evening is nearly as important for my workers as Christmas Eve is at home. We always stop work at three in the afternoon. The people go home, bathe, and meet an hour later in shiny, white clothes to receive an advance on their pay, one rupee. But the good Blacks cannot add, and therefore they always fight on Saturday evening. They always borrow money from each other, and, when they are to pay it back, they can barely spare it from the one rupee, and always want the creditor to wait until pay day. Since they are very good natured, they agree. But first they must make a disturbance. Outside the thorn hedge, which borders my garden, there are now two fighting about money. But, before I am finished writing this page, they will agree, but there might come others. It, however, disturbs their Saturday's pleasure so little that one does not notice. At the market place people buy maize *mtama*, sweet potatoes, *muhogo*, *mamaensje*, *mboya*, peas, beans and dessert is a huge amount of fruit. At sundown one sees them sitting in small groups around a huge plate of cooked vegetables. When they are finished with that, they begin on the fruit. They dance in the evening and the night until midnight, when all lay their heads down on their beds.

Yes! It is going well on the plantation now. I think that this year I will manage to earn 250,000 marks for the company. I have made my *Etat* [Budget] for the next year, ordered two new traction engines with 35 horsepower, and sixteen new machines to extract hemp with. I will then, next year, have here on the plantation 5 traction engines, 23 hemp machines, 21 brushing machines, 12 kilometres of field track, 32 wagons, 2 sailing ships, and, when I think that I experimented until 1899,[12] so, if I may say so myself, it is well done. Next year I will have a pure profit of 600,000 Mark. The company is nearly standing on its head. I sent the hemp to the exhibition in Dar es Salaam and received first prize, a beautiful bronze medal. I am very proud of it. I also have on the plantation 112 cows and calves, 32 oxen, who pull the hemp bales from the factory to the pier, 60 donkeys and 2 mules, which I ride all over the plantation. Some day you will get a picture of them.

Apropos! While I am writing I have come to think about my photographs at home. Please let me know that you are taking care of them. Remember, they are a part of myself, it is my life that lays in those two books. I would horribly regret losing just one of the pictures. Have Vilhelm put them in our box in the bank. There I know they will lie in peace. Order from bookbinder Christensen a book like the other two and send it to me. When I have filled it with pictures, I will send it home. Vilhelm asks me to send the numbers on my lottery tickets. How am I supposed to know them? I gave him the old notes, before I left and asked him to renew them for me

[12] Suggests that CL was working on ways to make sisal a commercially viable product from 1893 to 1899.

as long as I was in Africa. I have no idea what the numbers are. So, has he forgotten to renew them?

My plantation, and the land it belongs to, has a radius of 68 kilometres, and, if I use about two days to ride around the 7500 hectares, it is big. I sit like a little prince here in Africa.

Yes! I have written to Axel that he should go to America. I am convinced that he will do well there. But there isn't any real push in him. He cannot leave Denmark. It could never occur to me to come home and live in Denmark, if I did not have enough money to live on. I would a thousand times rather lay buried here in Africa. But I hope to earn good money. How is Musse in her hatbox?

Let me know, and write and tell me if Fusse passed her exam; I am interested. I do not believe in Sigrid.[13] Maybe I am wrong, but I don't think so. Yes! Hjalmar, how really dumb he was to marry at such a young age. That takes courage. I don't have enough courage now, and perhaps never will. It always hurts me to see a man work and struggle in his young years to support a family. It is, of course, wonderful and brave of him, when he does it. But youth, after all, is there to enjoy first, later it can be proper to marry when one feels alone and abandoned. First, when one, like myself, has reached the 40's, then it can be very good to marry. I want to also, but I know how I have it, and not how I can have it. Marriage is a dangerous lottery. Maybe one wins an angel or maybe a devil. When one has to fight with the woman every day, it won't be any honeymoon. So I prefer to live in peace with myself. But, if one could find an angel, a happy and sensible girl, yes, that would be another matter.

It is horribly unhealthy here this year. It is so cold that, while I am writing this, I have my summer coat on. We are all sick here on the plantation. I have lost twenty pounds, have had six fevers. My assistants are not doing too well either. But now it will soon be warm, so it will be better. I do not drink alcohol. Perhaps that is the reason I was so sick. But I wanted to see if one could hold out one year without drinking intoxicating drinks. It does not appear to have gone well.

Warm greetings to all
Your affectionate brother
Chr Lautherborn
Take care of the photographs

---

[13] Fusse must be a nickname, most probably for one of the many Carlsen children. Sigrid was one of Caroline and Vilhelm Carslsens daughters, born 1887.

Mwera, 12th February 1905

Dear Sister!
Many thanks for the letter, the pictures, and the almanac. Through them, one gets the impression that you have had a joyous and happy Christmas. Today is Sunday. All of my workers are in Pangani to buy food for the coming week. Here it is so quiet and peaceful, when I discount the large eagle floating over the house who, now and again, issues a scream, the peeping of the small birds and a couple of chickens, full of importance over laying an egg, which they announce with enthusiastic cackling. Otherwise one could believe that one was alone in the world, because there is no human sound heard; it is as if everyone was dead. Writing a letter during the week is nearly impossible, because, as soon as the Blacks know I am in the house, they always come and want to know about one thing or the other, and one is very disturbed. I also truly do not have much time for it. The huge upswing the plantation has taken always provides more work for the manager than for anyone else.

This year we spent our Christmas Eve on the little island Masive. When you look for it on a good African map, you will find it just out from the Pangani River. We sailed on our cargo ship (*Kinda*) at four from the plantation. Since we had to tack to get there, we did not reach it before six. We had sent our people in the morning, and tents were set up ready to receive us. We had a large dining tent and four, smaller ones to sleep in. We had three cooks, who were all enthusiastically occupied with preparing food. The German says, '*viele Köche verderben den Brei*' [too many cooks spoil the broth], but that was not the case here, because I have never eaten with more appetite than that evening. We also had, of course, a Christmas tree. The entire island is overgrown with *Kasuarins*, a tree which has much in common with fir trees. We decorated one of the smallest with gold thread, glass balls and silver balls, and when the candles were lit it looked enchanting. When we sang 'Stille Nacht, heilige Nacht' I followed the tune singing the Danish words 'Glade Jul, dejlige Jul' [Silent Night in Danish], without anyone noticing it. There is so little wind here at night (ordinarily), that the lights can burn under an open sky. The moon was so bright that lights were not necessary. But, for the sake of the holiday, we had four lamps on the table. Since the beach was filled with oysters, the goose was also naturally filled with them. We ate them raw and fried according to taste. Their reigned a wonderful, companionable atmosphere at the table, where we sat until nearly midnight. The talk was about the loved ones at home in the cold North. Everyone had to tell a couple of stories about Christmas in his youth. Every story recalled memories from the dear childhood years, and it was as if we were all children again. The atmosphere was nearly religious. Contributing to this was the wonderful weather and the magnificent place we had chosen for our Christmas Eve. The moon was high in the heaven, and the white sand under its beams looked like a smiling snow surface. The sea lapped with small, easy strokes on the beach, as if it slept and breathed weakly. The white houses in Pangani could be seen clearly. The Usambara Mountains, the small fort at the entrance to the Pangani River and the sea, the great, wonderful sea, had a silver sheen under the moon's light. Everything lay so peaceful and quiet, that it just had to affect every person open to the beauties of nature. Before we settled in, we took a bath in the clear, silver-shining water, came up on the beach, wrapped up in our blankets and slept so well that the sun began to shine on our faces before we awoke. Here is one place in Africa where one, without fear of wild animals, can sleep in the open, because there are none, and the island is so far from the coast that they cannot get over.

After drinking coffee we left the island and sailed back to Pangani. There we met all the Europeans from the different plantations. All the newest questions were discussed, and Christmas day passed wonderfully and joyfully. The next day we returned to the plantation; and Christmas ended. The children again took their books and went to school happy and satisfied.

That is also how it was with us. We were happy about having Christmas and about spending it so well. We allowed the tree to stand with all of its glitter to the great amazement of all the fishermen, who found it in the middle of the woods on the island and did not know what it was. But we have explained it, and now they know.

How is it with Lilly? Does she have a baby? I did not know anything about it. Gunnar will be all right.[14] He is young yet, but it will come. I think that especially military service will help him well on the way. At least he will learn to walk. I have heard from Cathinka that it is going well with Hjalmar. Well! He also deserves it. I am very pleased with the pictures. Do not forget to send me two books like the last ones. I will send them, filled, home. Do not forget to send me My System.[15]

Heartfelt greetings to large and small
Your affectionate brother
Chr Lautherborn

---

[14] Lily was the daughther of CL's youngest sister Cathinka, Gunnar Carlsen, born 1885, was a son of Caroline and Vilhelm Carlsen.

[15] Written with capital letters which might suggest it is the title of a book. The 'system' referred to may be an exercise system. The so-called 'Swedish exercises' were appreciated by progressive circles and people belonging to the Folk High School movement around 1900 in Denmark, so maybe it is an exercise book, perhaps with 'Swedish exercises' CL is referring to.

German East Africa Company
Pangani, 25th June 1905

Dear Vilhelm
I am in Pangani today. I am sitting in Surveyor Kayser's room, from which I have a good view over the river to my workers on the other side. Even though it is Sunday, we have to work. The hemp cultivation has gone so well that we, with our little schooner, can no longer transport the large amount of hemp from here to Tanga. A steamship from the East African Line lies in the middle of the river, and, with small boats, we are bringing the bales from the beach to the ship. The sun is high and clear in the sky and sends its rays with strength down here bringing out the sweat on the Black's faces and bodies. It is like when we take a steam bath at home, but it does not affect their cheerfulness. I can still hear their laughter and yelling, it is, in any case, evidence that they are not sleeping.

I have recently added a new machine, which the Blacks consider to be one of the seven wonders. It is a highway locomotive with 35 centimetres wide wheels. Even though it has a weight of 22,000 pund [pounds], I can drive it in quite soft earth and pull 16,000 to 20,000 pund [pounds] of hemp. It greatly impresses the Blacks. When I, the other day, drove it down the highway, that goes from Bweni to Kikogwe, and which raises 1 fod [foot] in 10, I had at least 2000 people behind me. All were excited over the fact that the machine could drive on so steep a road and thought it could climb like a goat.

You must not believe that I, today, on a holy day, will write something for *Vendsyssel Tidende*, but I found an article, which I have known about for several years. It was told me in confidence, so I have, naturally, said nothing about it. I knew Lieutenant Langheld very well in the years '90, '91. He is the type of person one likes after a short acquaintance. He was very moderate with drink, and was lucky enough to nearly ever have fevers. When it happened, he was only a little weakened. I was also a little acquainted with Mr. Werner. We were both with the same company. The Sudanese are the bravest blacks I have known in my life, and I have never noticed that they were superstitious. Have my article (*Mein Freund Werner Tod*) [My Friend Werner Tod] translated and give it to the dear Vendelboere [the people in North Jutland, that is the ones that read *Vendsyssel Tidende*] with the remark that I believe it.

Greetings to Caroline and the children and the family and good friends.
From your affectionate brother-in-law
Chr Lautherborn
As I said, it is Sunday, and I have no glasses here in Pangani.

Mwera, 15th October 1905

Dear Sister!
Thank you for your letter. Thanks for the picture of Father's and Mother's grave. Thanks for my system, and thanks for everything good. I should have long-since answered you, but I have been so busy that I, beyond my work, have hardly any time for anything. Recently two directors and two inspectors were here to see the plantation, inspect the books etc. They were so happy and satisfied with my work here that they promised me a 2000 Mk. bonus and the large gold medal. The money I get immediately, the gold medal when I leave the company. This year we will earn 2 million marks. I have just finished a rough estimate of what the yield will be from planting 200,000 *Kautchuck* [rubber] trees. In three years one can have earnings of 100,000 marks.

I am seriously speculating about bringing Axel and Gunner down here with me. Home in Denmark they can continue until they are old and grey and very likely not earn much money; on the other hand, if they come here, they can, in the course of eight years, return home as rich men with at least 300,000 Kr. in their pockets earned by planting *Kautchuck*. But, when I take on the responsibility of taking two, young people with me down here, it must naturally be completely stable and safe in the country, but that is not how it is here. Down in the south the Blacks are killing the Whites. A good acquaintance of mine, a Herr Hopfer, fell as the first victim. Now the uprising has spread over half of the colony, and here unrest is growing steadily. I keep my rifle in order, and two pistols lie loaded on my bedside table, because one can never know when the *Zigua*[16] will want to kill us and steal our property. However I do not think the uprising will last long, because the Blacks have very little powder and shot, and it has been initiated by a number of witch doctors. They have convinced the superstitious Blacks that they are connected to a great snake, a god called Kola. He has given them the power to shield any man's life and the power, with the holy water in which the snake lives, to make any man bullet-proof. To put it more precisely, if the government soldiers shoot at any man washed in the holy water, water will come out of the gun instead of shot. The Blacks have bought the water and have such strong faith in it that they go against the soldiers' guns unarmed. Whether or not they will kill the witch doctors, when they are, naturally, mowed down like grain, and they discover their magic doesn't work, is a big question.[17]

---

[16] The *Ziguas* were one of the key people in the Maji Maji revolt. They were based in the lower Pangani river basin. See chapter two.

[17] *Maji maji*, an uprising of African peoples against German rule, erupted in the fall of 1905, but was not active near Pangani for the most part. The uprising started at a cotton-plantation where some Africans had torn up cotton plants as a protest against the working

Here, to date, we have been spared from the war, and it would truly please me if we could escape it. There is, however, an enemy just as bad, i.e. the plague, now in Zanzibar, and I do not believe it will be long before we have it here. Two Europeans are already dead of it in Zanzibar, but we hope for the best, that we do not get it here. You must not say anything to Gunner or Aksel, because I do not know if they can come down here with me. I will think about that until I come home in 1907. It is often the result when one promises a young man something that he lives in expectation and forgets to pay attention to what he is doing. That Kirsten and little Ellen were engaged with the brothers Herfort I have known long.

That Kirsten was secretly engaged with the doctor, or that they had an especially good eye for each other, I could see when I was in Copenhagen. You remember that evening you had such a terrible headache and Cathinka had to bring you home. Well! So, you have only one grown daughter left, Sigrid, but I do not believe it will be long before she has found herself a man. So Frode wants to be a farmer.[18] Yes, that can, of course, be very good. It is, in any case, the best that I know, but one has to be lucky; otherwise it is a life full of drudgery with only a little profit. I had thought that Frode was to study to be an editor, so that he could take over the newspaper from Vilhelm when he was ready to stop, but I now see that that was a mistake.

I would like to have Aksel where he is earning something. Not because I am bothered by the money he gets from me, but because it will make a better man of him to come so far. I earn so much that I can pay out and still lay something aside. He has, after all, to this date, not earned one red øre [never earned just one dime]. It would, therefore, please me to hear that he had now come so far that he could support himself. I do not think it could hurt him any bit to take a position as a barber's apprentice, until he has thoroughly learned the profession. There is time enough for him to get something for himself.

I have today, Sunday, sat at the desk the entire day. The clock now shows 4:15 in the afternoon, and I would like to take a tour down to the factory

---

conditions on the plantation. The Germans' harsh reaction just strengthened the Africans resolve to fight the Germans. 'Maji' means water in Swahili and refers to the holy water supposed to protect the African warriors as described in CL's letters. The uprising—or war—did not, however, come to a quick end. It lasted for two years and did cost around 100,000 African lives whereas only 15 English were killed. (Felicitas Becker, Jigal Beez (eds.), *Der Maji-Maji-Krieg in Deutsch-Ostafrika 1905-1907,* Christoph Links Verlag, Berlin, 2005). The uprising convinced most colonial officials that the German colonial state could not coerce sufficient African labour to make plantation production economically viable. After the uprising ended in 1906, the administration moved slowly and cautiously towards African small- scale agriculture and against plantations (Sunseri 2002), (see also chapter two).

[18] Frode Carlsen, son of Caroline and Vilhelm Carlsen, born 1889.

to see if the hemp has been stored correctly. You will, therefore, have to excuse me for now, and I ask you to give my regards to the entire family.

From your devoted brother
Chr. Lautherborn

Mwera, 5th March 1906

Dear Little Regine[19]
Now I have received two letters from you without answering them, but today it will be done. First let me tell you that the lions we have here are not in possession of a mane, but its skin looks like cowhide. On the other hand we have beautiful panther skins, and what would you say to a zebra skin? I believe it would look especially good in a parlour. Now I will see what I can do for you. Now, with regards to the carpets, I will see if I can find a couple of ten and eight alen size ones. I believe it will cost about 600 kr. for genuine Persian or Smyrna. Imitation goods are naturally much cheaper and will run up to a value of 150 to 200 kr, but they do not look very good.

Unfortunately we have, at this time, plague in Zanzibar, and, since that is the only place I can get the carpets from, you will have to wait until the plague has gone. I was very saddened to hear that Hjalmar was sick of diphtheria, but I was equally glad to hear that he had gotten over the illness. You don't need to worry about heart paralysis with him. Marie has paralysed his heart and will continue to do so, every time she provides him with a little heir. I believe she has done this four times, so his heart must soon be immune. By the way it must be wonderful for Ellen to finally fulfil her obligation to her brother and, in addition, under the eyes of her Jonas; that is a fine combination of the useful and the pleasant.

Cannot you now tell me a little about Aksel. Has he become a barber's apprentice? Does he work in Copenhagen at the profession, or what is he doing? I certainly hope he has not become a burden for Cathinka. In your letter it stood that Ashe has had a bad man, and that is not good when one is young and beautiful. I do not completely understand that. I know it is not going so well for Emil as it could. I am very happy that I am not his child, I mean for my sake; I believe she is a real Salass [German word?]. I feel sorry for both of the girls, who are home; I would really like to marry both of them to get them out of the home and out among friendly people, but that will not do with civilized people. I have not seen Brussels carpets

[19] Regine must be Regine Carlsen, daughter of Caroline and Vilhelm Carlsen, born 1877, married 1899 to Niels Otto Heerfordt. Actually three of the Carlsen daughters married into the Heerfordt family.

here, but, when I go home, I can make inquiries in Berlin. I think one can get them cheap there.

The plague and the *Ziguas* have not harmed me. The rebellion is more in the interior, and the plague remains in Zanzibar. The plague has, however, come to Uganda from the English territory, where it has undoubtedly been brought in by the Indians. It has been followed by sleeping sickness, which is absolutely deadly. A little fly, like the one that gives the illness to cattle, introduce into people, through a sting, an amount of infusoria which sets itself in the spinal cord. The infected begin to sleep, wake up only to complain, get something to eat and sleep again. It continues in this manner until he has slept into eternity, actually a pleasant death.

Otherwise I am doing well. I work my system through every day and feel like I was eighteen. I eat very little and drink no alcohol when I am alone. I have now tried this for one year, and I feel wonderfully well. Previously I had always believed that it was necessary to drink spirits in the tropics, but I have now come to a completely different opinion.

Farewell for now.
Greetings to all
Your Chr Lautherborn

[This letter was written by Christian Lautherborn's assistant, Reicke, to Caroline Carlsen]

Kikogwe, 4th June 1907

Most honourable Madam,
First of all excuse me because I have not answered your friendly letter that reached me at the end of February. I assure you that it was not caused by forgetfulness. Unfortunately I have been ill for a long time and afterwards had pressing duties to attend to, therefore I have not been able to answer your letter before now.

Madam, you do not need to thank me. What I did for your late brother in the last days of his life was not only a duty, which in this case anyway was a duty dear to my heart as a better and more friendly superior one could never hope for, it was also a repayment for the many times your brother assisted me during severe illnesses. We all did our best, but nevertheless it was not in our power to avert God's will.

I will oblige your wish and describe his last days to you in some detail. Already at the beginning of September he complained now and then of feeling less well with fevers and pain in between. He was advised by several people to expedite his vacation planned to start in May 1907. However, he did not listen to this advice and his general condition deteriorated from day to day. In between he had days where he felt quite well, other days he had to stay indoors, the severeness of the fever had taken away his strength.

Sunday, 4th November [1906], I had to go to Tanga on business. When I returned Wednesday, 7th November, I found that your honourable brother had caught the black fever. His condition worsened Thursday, and Friday it became clear that the end was drawing near. Saturday morning at 5 o'clock I rode over to the factory after having attended to your brother, whom I left sound asleep. When I returned at half past seven, I noticed that his breath was oppressed and that in between he had difficulty breathing at all; he was also beginning to lose his consciousness.

I rode off as fast I could to get the doctor in Pangani and he returned together with me as fast as we could. The doctor stated that the patient due to loss of blood the previous days was severely enfeebled. This enfeebled condition lasted from midnight until 9 o'clock the next morning; the patient lay unconscious in between, but other times he woke up again. In his clear moments he asked questions about plantation management and expressed a wish to be carried over to his own house in Mwera (he died in my house in Kikogwe because he suddenly fell ill here). He also once expressed his wish to write a letter to you Madam, stating that he would soon come to visit you. He was, however, not capable of writing any longer as he had lost all his strength. At 9 o'clock the same evening, he lost consciousness and ten minutes before midnight he passed away. I do not think that he was truly aware at any time, or admitted to himself, that he was dying. To the end he kept up hope and planned to go to Tanga on the next steamer and go to the hospital there.

In order to oblige your wish to know how the grave has been laid out I enclose the attached photos. The grave lies beautifully framed by three baobab trees close to the sea. Under the tree in the middle is the grave. In the background you can see my house where he died, the small window is the window in the room where he died.

It will be a dear duty to me to keep his grave well maintained. I assure you that I will never forget your late brother, who during his industrious life accomplished so many successes. I will always keep him in my memory.

Be assured, Madam, of my best wishes and highest esteem.
Yours respectfully,

Reicke

BIBLIOGRAPHY (FOR CHAPTERS 3-9)

Becker, Felicitas and Jigal Beez (eds.) (2005) *Der Maji-Maji-Krieg in Deutsch-Ostafrika 1905-1907*. Berlin: Christoph Links Verlag.

Beidelman, T. O. (1982) *Colonial Evangelism: a socio-historical study of an East African mission at the grassroots*. Bloomington, Indiana: Indiana University Press.

Bennett, Norman R. (1986) *Arab versus European: Diplomacy and War in Nineteenth Century Africa*. New York: Holmes and Meier.

Brode, H. (1911) *British and German East Africa: Their Economic and Commercial Relations*. London: Arnold Publishing.

Brown, Walter (1975) "A Pre-colonial History of Bagamoyo: aspects of the growth of an East African coastal town." PhD Dissertation, Boston University.

Donzel, E. Van (ed.) (1993) *An Arabian Princess between two Worlds*. Leiden: E. J. Brill.

Gilbert, Eric (2004) *Dhows and the Colonial Economy of Zanzibar*. Columbus: Ohio, Ohio University Press

Glassman, Jonathan (1995) *Feasts and Riot: Revelry, Rebellion and Popular Consciousness on the Swahili Coast, 1856-1888*. Portsmouth: N.H. Heinemann Press.

Iliffe, John (1979) *A Modern History of Tanganyika*. Cambridge University Press.

Jespersen, Knud J. V. (2004) *A History of Denmark*. Palgrave/Macmillan.

Koponen, Juhani (1994) *Development for Exploitation: German Colonial Policies in Mainland Tanzania, 1884-1914*. Helsinki: Finnish Historical Society.

Kurtz, Laura S. (1978) *Historical Dictionary of Tanzania*. Metuchen, N.J.: Scarecrow Press.

Mann, Erick (2002) *Mikono ya damu: "Hands of Blood:" African mercenaries and the Politics of Conflict in German East Africa, 1888-1904*. Frankfurt: P. Lang.

Ofcansky, T. P. and R.Yeager (1997) *Historical Dictionary of Tanzania*. Lanham, MD: Scarecrow Press.

Rockel, Stephen (2000) "'A Nation of Porters': the Nyamwezi and the Labour Market in Nineteenth Century Tanzania," *Journal of African History* 41, 2.

Sicard, S. von (1970) *The Lutheran Church on the Coast of Tanzania 1887-1914*. Lund, Sweden: Gleerup.

Sunseri, Thaddeus (2002) *Vilimani: Labour Migration and Rural Change in Early Colonial Tanzania*. Portsmouth, N.H.: Heinemann Press.

Sutton, J. E. G. (1970) *Dar es Salaam*. London.

GLOSSARY

Bagamoyo: a prosperous trading port on the East African coast where Lautherborn spent two years during the East African War

Bweni: a small town across the river from Pangani

Dhow: an African sailing vessel with one large sail

DOAG: The German East Africa Company

GEA: German East Africa

Grundtvigian Movement: Revivalist movement within the Danish Church with N.F.S. Grundtvig as its ideological inspirator

Kikogwe: Lautherborn's first and primary plantation near Pangani

Hjørring: the town in North Jutland, Denmark, where Lautherborn's brother-in-law ran a newspaper. The Hjørring Museum held Lautherborn's letters, photos and artifacts.

Holy Ghost Fathers: the Roman Catholic mission near Bagamoyo.

Home Mission: Revivalist movement within the Danish Church, in opposition till the Grundtvigian Movement

Majumbe: African or Shirazi chiefs (singular is *jumbe*)

Maliwali: the Arab representatives in the coastal towns (singular is *liwali*)

Mrima: the East African coast opposite Zanzibar

Mwera: Lautherborn's second plantation, also near Pangani

Nyamwezi: the peoples from Unyamwezi in central GEA, who were active in the caravan trade

Omani Arabs: the ruling elite based in Zanzibar but originally from Oman

Pangani: a coastal town North of Bagamoyo, with a long history of coastal trade and slave production of sugar and other products.

Vendsyssel Tidende: the newspaper of Hjørring, edited by Lautherborn's brother-in-law, Vilhelm Carlsen.

Zanzibar: an island off the coast of GEA, where the Omani Arabs had their headquarters from which they controlled much of the trade and production of the mainland.

INDEX

SOURCES FOR AFRICAN HISTORY

1. J. Silvester and J.-B. Gewald. *Words Cannot be Found.* German Colonial Rule in Namibia: An Annotated Reprint of the 1918 Blue Book. 2003. ISBN 90 04 12981 2
2. J.B. Lockhart and P.E. Lovejoy (eds.). *Hugh Clapperton into the Interior of Africa.* Records of the Second Expedition 1825-1827. 2005. ISBN 90 04 14155 3
3. M.W. Daly and Jane R. Hogan. *Images of Empire.* Photographic Sources for the British in the Sudan. 2005. ISBN 90 04 14627 X
4. Jane L. Parpart and Marianne Rostgaard. *The Practical Imperialist.* Letters from a Danish Planter in German East Africa 1888-1906. 2006. ISBN 90 04 14742 X